Conten

Foreword

This is the first of three pupil's books specifically written to cover the Junior Secondary School Agriculture syllabus for Sierra Leone. The books have been written by three authors from Sierra Leone and one author with many years experience in the country, with the objective of producing textbooks that are specifically relevant to the Sierra Leone situation.

The books have been written to allow pupils direct involvement with the work. The style throughout, using questions, activities and exercises, as well as relevant explanations, will help to develop pupils' sense of enquiry, and their ability to observe and record information. In this way they will be encouraged to adopt a scientific approach to agriculture.

The three books follow the spiral nature of the syllabus, whereby themes introduced in Book 1 are developed and explored in subsequent books. All the essential areas of agriculture are covered so that pupils will be better able to practise agriculture at the end of Junior Secondary school, or continue on to study for the West African Examination Council's Senior Secondary examinations in Agricultural Science under the proposed 6–3–3–4 educational system.

Each chapter deals with a specific topic and the teacher can choose to follow a sequence to suit his or her teaching programme. Factual information contained in each chapter may be used by the teacher to lead the class, or the pupils can read on their own. This can be followed by practical activities, question and answer sessions, discussions, puzzles and field visits which provide interesting variety.

Agriculture is the primary industry in Sierra Leone and the Government is encouraging increased agricultural productivity. It is the authors' intention that these books will make a major contribution to this effort by helping to train future generations of farmers and agriculturalists. The emphasis in this course, therefore, is on the relevance of agriculture to the way of life and economy of Sierra Leone. Pupils are encouraged to realise that employment in agriculture can be both prestigious and rewarding. Agriculture must assume greater importance if the country is to achieve self-sufficiency in food production and reduce its dependence on imported foodstuffs which utilise vital foreign exchange.

The books recognise the important contribution made to the national economy by both traditional agriculture, as practised by local village communities, and modern scientific farming; the books encourage pupils to understand both systems. While the authors seek to promote increased agricultural production, the conservation of the nation's resources and the development of sustainable systems of farming are key themes throughout the books.

Mr Josephus O'Reilly
Mrs Kenye Turay
Mr Michael Kabia
Mr Ralph Liney

Acknowledgements

The authors and publishers would like to thank Dr Sonny Tucker of Njala University for providing useful comments on the manuscript.

CHAPTER 1
An introduction to the agriculture of Sierra Leone

Introduction

Welcome to the first of three textbooks written specially for you to study agriculture. You will discover how important it is that everyone in Sierra Leone understands what agriculture is all about!

In this chapter you will learn what we mean by agriculture and farming, how agriculture developed from the earliest times, and what farmers do today. You will learn about the importance of farming to our country and will understand how farmers can play a part in looking after our forests and wildlife. We hope you will enjoy studying this important subject.

The meaning and definition of agriculture

If we look carefully at the various activities in which men and women take part in order to feed themselves and to provide food for others we shall have a picture of what agriculture and farming are all about.

ACTIVITY 1

Have a look at the photographs in Figure 1.1 and you will see that men and women, who we call farmers, are doing agricultural work in a number of situations. Write down the numbers of the pictures in your book and explain in a few words what you think they are doing. You will probably know some of the activities, but possibly not others. Ask your teacher after you have finished to explain them.

The main function of agriculture is to provide food for humans. Some farmers use their knowledge to grow plants (called crops) which they harvest as the first stage in this process. Other farmers keep animals (called livestock) from which milk, meat and other products are obtained. When a farmer produces more food than is required for the family, it can be offered for sale (in return for money) or barter (in return for goods or services). Can you think of some things that farmers commonly offer for sale or barter?

1

2

3

4

5

6

Figure 1.1
What are the people in the pictures opposite doing?

Some agricultural products may be processed into goods; for example, milk may be made into butter or grain may be made into beer. This business of processing or manufacture makes the farmers' goods more valuable. Can you name three agricultural products that are normally processed by farmers before they are used?

Now we can bring these ideas together and say that agriculture may be defined as the growing of crops and the rearing of livestock to feed us and others and to provide goods for the benefit of us all.

We should remember, however, that farmers in Sierra Leone engage in many activities in the countryside which fall outside this rather strict definition. We must not overlook the importance of such activities as wood gathering, fishing, hunting and the harvesting of forest products in providing for the needs of the family.

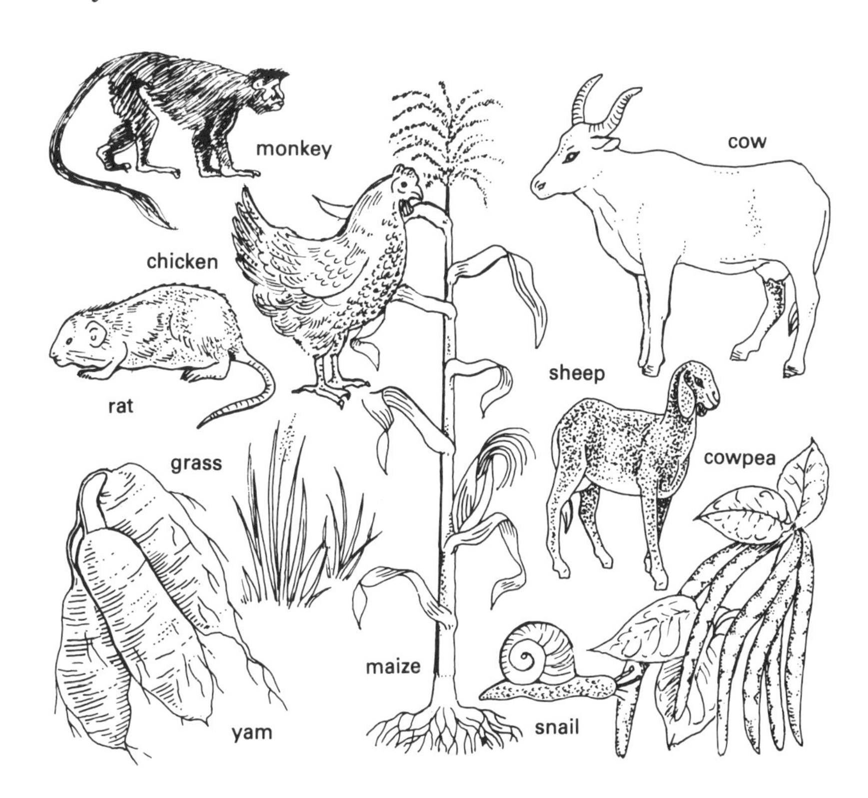

Figure 1.2
Which of these plants and animals are useful and which are wild?

ACTIVITY 2

Examine the pictures in Figure 1.2 and decide which plants and animals are useful to us and which are wild or may do damage to our crops and food supplies. Remember that wild animals are part of what we call nature and need food themselves. You may drive them away, of course, but do not kill them unnecessarily unless they are a danger to you, your crops or your livestock. Now draw four columns in your notebook, as you see in Table 1.1, with

Table 1.1 Plants and animals

Plants		Animals	
Crops	Wild plants	Livestock	Wild animals

the same headings. Write down your answers in the columns and when you have finished compare with others in your class. How many did you get right?

History and development of agriculture

What did early people do?

Agriculture started almost as soon as the earliest people lived on Earth. We can therefore say that the development of agriculture is closely related to that of the human race.

Early people must have lived at least one million years ago. They were smaller than we are today but they stood upright and walked like all of us. They needed food but did not know how to keep animals or how to grow crops, so wild animals were hunted to provide meat, while fruits, berries, and roots were gathered in the forests. At that time, early people were wanderers or hunter/gatherers. They killed animals using spears made from sharpened stones attached to straight sticks and clubs made from strong tree branches. During this period, the countryside was very different from today and the small number of people living in an area would have existed as small family groups. All family members would have helped with the work of finding seeds and fruits to eat; this probably would have included wild rice which would have been found in the wet areas. Later still, people found out that, by rubbing stones together, they could produce fire on which to roast meat and to cook. In this way they survived and gradually developed in understanding and practical ability.

What was the countryside like then?

The high land would have been rocky and bare with few trees and would have been quite cool. At lower levels the heavy rainfall and higher temperature encouraged forests to grow, making jungles of tall trees and climbing vines. The ground would have been thick with roots and fallen leaves. This was where many wild animals lived. In places, the forest gave way to the open plains we call savannas, which provided grazing for large herds of animals; these animals were hunted by meat-eating animals and people (Figure 1.3).

Figure 1.3 Early people hunting and gathering

Living together

As time passed, people found that living together in larger groups of several families was better for everybody. They were able to work in teams; the men hunted in parties, while women and children searched for fruits and berries, made clothes and prepared food. As there were more people, work could be split up, making it easier; some people with special talents would have been able to develop their skills (people might, for example, have become skilled at pot making).

Living in larger groups gave added protection as people could defend themselves from wild animals and from their enemies; in this way the groups settled in villages. These settlements, made up of huts, were built in convenient areas and the land nearby was gradually used for growing edible plants. Slowly certain wild animals were tamed and were used by people. This process, which took place over many thousands of years, provided the basic structure for our agriculture and society as we know it today.

How was progress made during that period?

As we have seen, people found that seeds of plants could be made to grow, so seeds were collected from wild plants and specially grown on land near to the village. They found that trees could also be grown from seeds found in the fruit that they had gathered and this helped to feed the family. Palm fronds were found to make a good roofing material. Certain muds could be shaped into pots and dried in the sun; later it was discovered that they could be made much stronger by heating them in a fire. Dried mud blocks were used for making huts. It was found that wild cotton and other fibres made rough cloth and rope and that branches from trees made posts for house building.

Figure 1.4
An early settlement

Later, people discovered that some animals could be tamed to live near the village or even in the huts or yards (Figure 1.4). Dogs were probably the first to work with people, but then cattle, sheep and goats were tamed and bred. They provided meat, milk, hides and wool. So progress was made, but slowly.

Markets

Men and women growing crops and raising livestock (who we now call farmers) began to produce more than they needed for themselves. In order to get rid of these extra goods a form of barter started in which people exchanged surplus produce. This gradually gave way to a system of exchanging food and materials for money.

Figure 1.5
Vegetables being sold in a market in Freetown

People often gather in groups in selected areas to sell their goods; such areas are known as markets. It is in the marketplace that the sellers and buyers of produce agree the price of the things for sale (Figure 1.5).

Have you ever given a friend something you owned in return for something that you wanted? How did you arrange a fair exchange and make sure you were both satisfied?

ACTIVITY 3

This activity is a class discussion which your teacher will lead. Many of the questions you will discuss will require you to think quite a lot!

Perhaps there is an agricultural market close to you. Why is that market found there and why do people gather at that place to sell their goods? What goods are offered for sale? Do the type of goods on sale change with the season? How do the buyers and sellers agree on a price for goods? Is there always a lot of argument at the market? Why is this? What things affect the price a farmer may get for his goods?

What does the farmer grow

In Figure 1.6 you will see examples of some of the many crops that farmers grow in Sierra Leone. In order to grow these crops the farmers must know, for example, when to plant the crops, how to plant them and when to harvest. People who have spent their lives farming have worked hard, have developed a great deal of knowledge and should be respected. You will have seen men and women going to their farms in the morning, carrying implements and food, probably to stay out for the whole day. Have you ever

Figure 1.6
Some of the crops grown in Sierra Leone

wondered what they were going to do? Perhaps you have helped in the fields. In this section we will look at the type of work farmers carry out in the fields during the year (Figure 1.7).

At the start of the rains, or just before, farmers cut down dried crop stems and weeds, pile them into heaps and burn them. Then, using hoes (or possibly ox ploughs) they turn over the top layer of the soil and break down the surface into small pieces suitable for planting; this cultivated soil is called a tilth.

When the soil is damp, the planting or sowing season can start. Depending on the crops to be grown, farmers may plant seeds, which are then covered with soil, or they may use planting material (with or without roots) which is placed in the soil at intervals. Alternatively, rice seedlings, previously grown in nursery beds, may be pulled up and planted out in the wet lands or swamps; this is called transplanting. Can you name four crops that are grown from seeds and four crops that are grown from planting material?

(a)

(b)

(c)

(d)

Figure 1.7
Farming activities:
(a) harvesting
(b) spraying
(c) weeding
(d) planting

During the growing season the crops must be kept free from weeds (unwanted plants). You have probably helped with this work; it is a job which young people do very well!

A farmer may then sprinkle fertiliser on the crops; this can be done by hand or with a machine. Fertiliser helps crops to produce good yields. Farmers may also use chemical sprays to control insects, pests or diseases. In the swamps, farmers may have to control the water supply to the rice crop; this is called irrigation.

As the crops mature, birds may have to be kept away. Have you ever helped with this? Which crops do birds like to eat most?

At the end of the growing season, crops ripen and harvesting can begin. Root crops are dug up and grain crops are beaten to remove the grain from the heads of the harvested plants before drying. Some of the harvested crops will be eaten by the farmer and family, some will be stored, and any extra can be taken to the market for sale. Now cattle and goats may be allowed on to the land to eat any fallen grain or other edible material; the land is rested to be ready for another year.

ACTIVITY 4

We have now looked at the farming year and in Figure 1.8 we can see a chart of the activities that take place on a typical farm in Sierra Leone. You should record in your own book what is taking place on a local farm. Copy the chart shown in Figure 1.9 into your book; don't forget to draw the picture of the farmer in the centre! Ask your teacher to draw a similar diagram on a big sheet of paper that can be put up on the classroom wall. During the year, record in your own book what is happening on a local farm (your teacher will help your class fill in the chart on the wall). When you have finished you will be able to tell what you might expect to see on that farm at any period in the year. At each stage in the farming year you should discuss what is happening with the rest of your class.

Figure 1.8 Some of the activities in a typical farming year

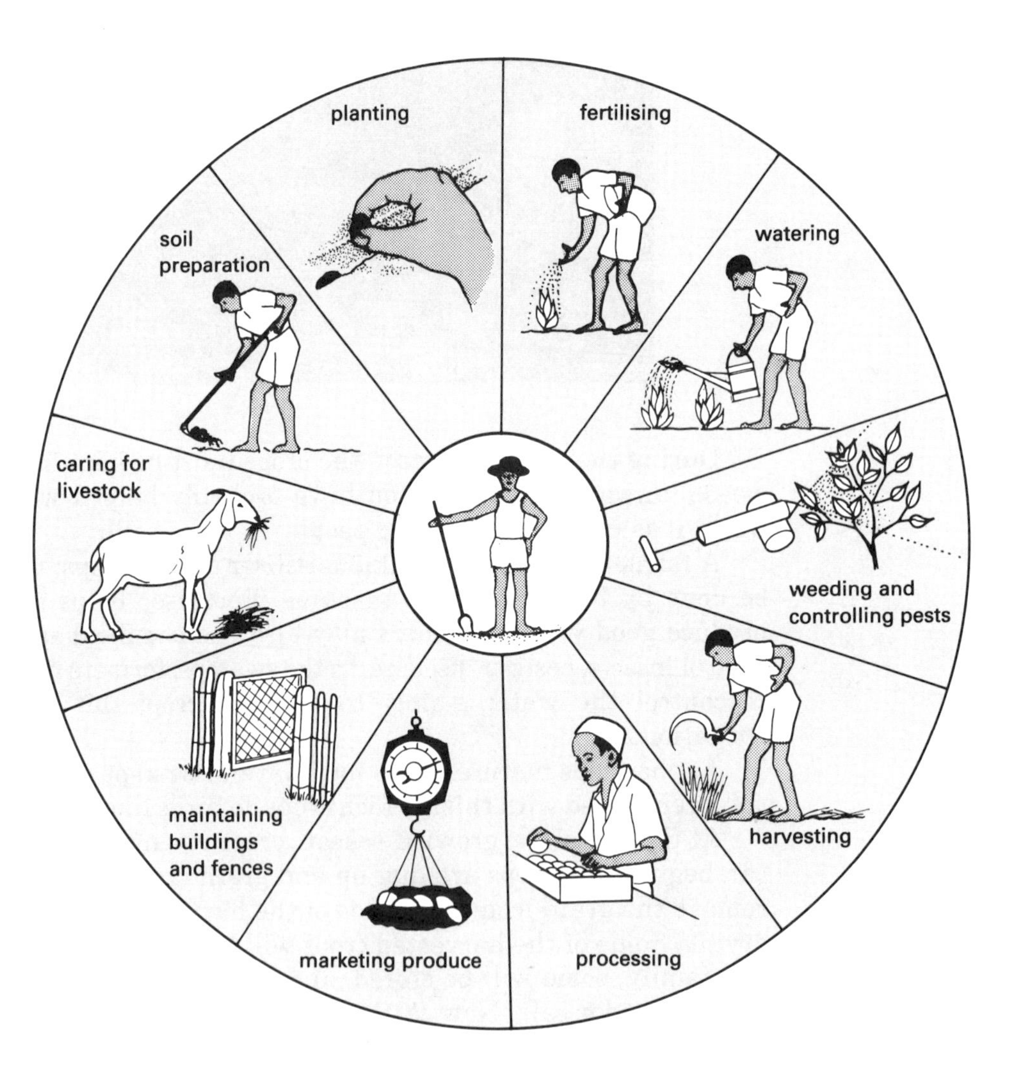

Figure 1.9
Copy this chart into your notebook

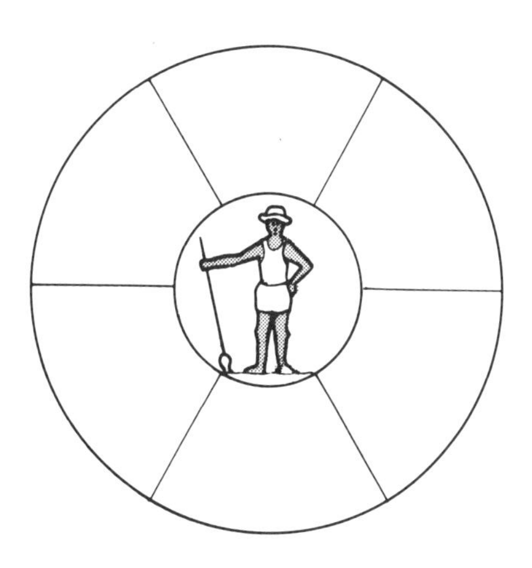

Livestock

In thinking about the farming year we must not forget the important place of livestock in the agriculture of Sierra Leone. In Sierra Leone there are a few very big farms that keep large numbers of livestock and in the north there are people who keep large herds of cattle. But most animals in the country are kept in small numbers by their owners close to their villages. Sometimes livestock can be a nuisance to farmers who are trying to grow crops as the animals may break into their fields and gardens and eat or trample the plants. However, cattle and other livestock, if properly controlled, can be reared successfully alongside the production of crops.

The cattle found in Sierra Leone are able to live very well in our climate and make good use of many different forms of vegetation. These animals eat grass, leaves and crop remains and, as they are moved from place to place, they leave manure on the soil which improves the soil quality for next season's crops. Sheep, goats, pigs and poultry also return manure to the soil and they too must be controlled. So you can see that if livestock are properly managed they can help in crop production! Who owns the farm animals that are kept close to your home? How are they prevented from destroying other farmers' crops? What happens if one person's cattle eats another farmer's crops? Discuss these questions with other pupils and your teacher.

How does the farmer know what to do?

We have already seen that farmers who have grown crops for most of their lives have a great deal of knowledge. How did they get this knowledge and learn what to do?

When today's farmers were young they would have watched their parents working on the land and would have helped them, just as you may do now! In this way they learnt the knowledge and developed the skills needed to grow crops and keep animals. In time, when they grew up, they carried on in much the same way as their parents.

Science has recently come to the farmers' aid and new farming methods and practices have been developed. For example, fertilisers and pesticides make life easier for farmers and machines can help farmers to grow much more than in the past. All these techniques, as they are called, help to increase the amount produced. Farmers must learn about these techniques if they are to progress; this is why we are studying agriculture in school and why the Ministry of Agriculture gives farmers free advice.

We can say, therefore, that a successful farmer is one who makes use of his or her knowledge of local conditions and uses scientific improvements to produce food for the family and sell food to others in the community. Such a farmer will make a good contribution to the economy of Sierra Leone. Do you know any farmers? Which of these farmers is the best? Why do you think that this farmer is better than all the others? Write his or her name in your exercise book and say why you think he or she is the best.

The importance of agriculture

Agriculture plays a key role in the life and development of a nation, especially those countries whose economies depend largely on agriculture. Sierra Leone, Guinea, Ghana, Nigeria, and Gambia are examples of countries that depend heavily on farming for their prosperity.

Also, because the population of Sierra Leone is growing very quickly, there is an increasing demand for food and other agricultural products. Agriculture is therefore of vital importance to our country if everyone is to enjoy a healthy life and a good standard of living (Figure 1.10).

Farming is important in Sierra Leone and other developing countries because agriculture provides:

- food for healthy and growing populations
- income and employment for many people
- goods for export which earn foreign exchange
- drugs and herbs to keep people healthy
- raw materials for manufacturing industries
- feed for livestock
- building materials for house construction
- materials for making clothes

Can you think of examples of agricultural products which apply to each of these groups?

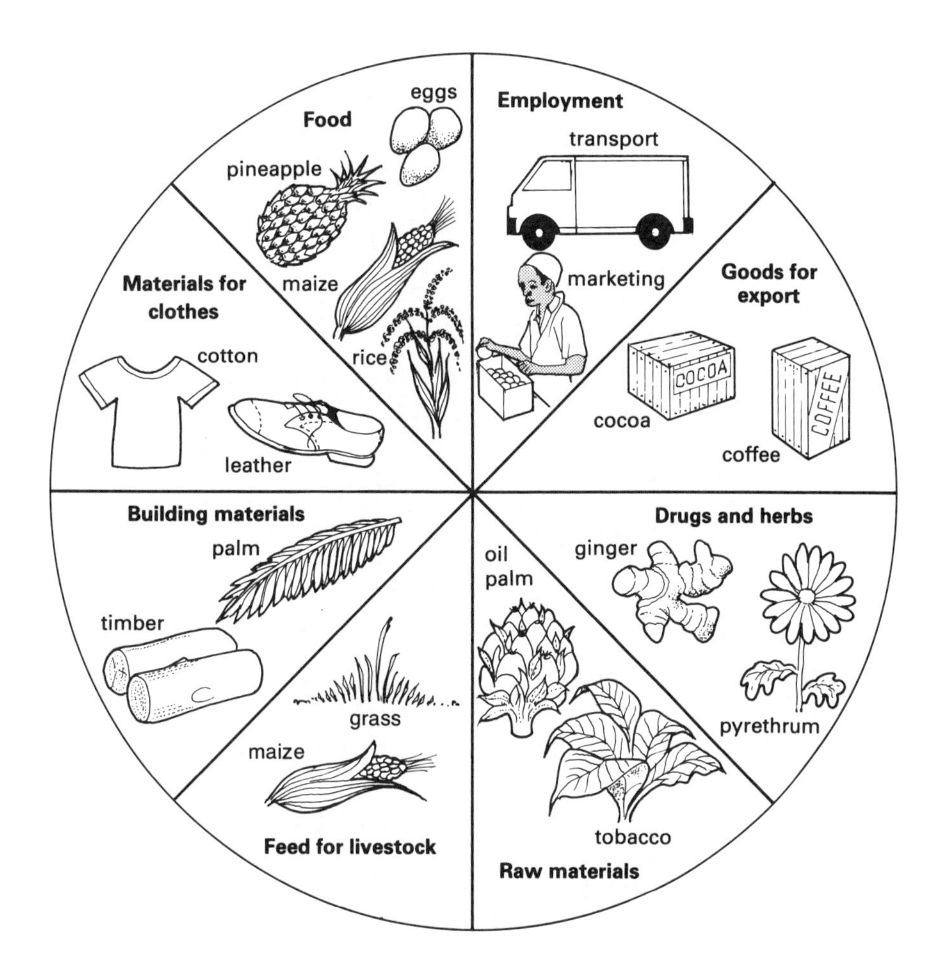

Figure 1.10 Examples of agricultural products to show the importance of agriculture in Sierra Leone

Let us now consider each one of these points in more detail.

Providing food

Agriculture provides all the food items that we consume. We have seen that crops and crop products are produced by farmers; these include rice, sorrel, krain krain, potatoes, cassava, yams, ginger, pepper and onions. We must also remember the wide range of crops that provide us with delicious fruits which add variety to our diet; pineapple, mango, guava and pawpaw are only a few examples. Can you think of some more? Farmers also produce livestock products such as eggs, chicken meat, beef, mutton, and rabbit meat. Food is the single most important item produced in our country and without it, the population cannot live.

Farmers must produce the correct type of foods as everybody needs a mixture of foodstuffs to remain healthy; this is called a balanced diet. Different foods provide different parts of a balanced diet. We get protein from the meat we eat and fats and oils from palm oil, which is extracted from oil palm nuts (Figure 1.11).

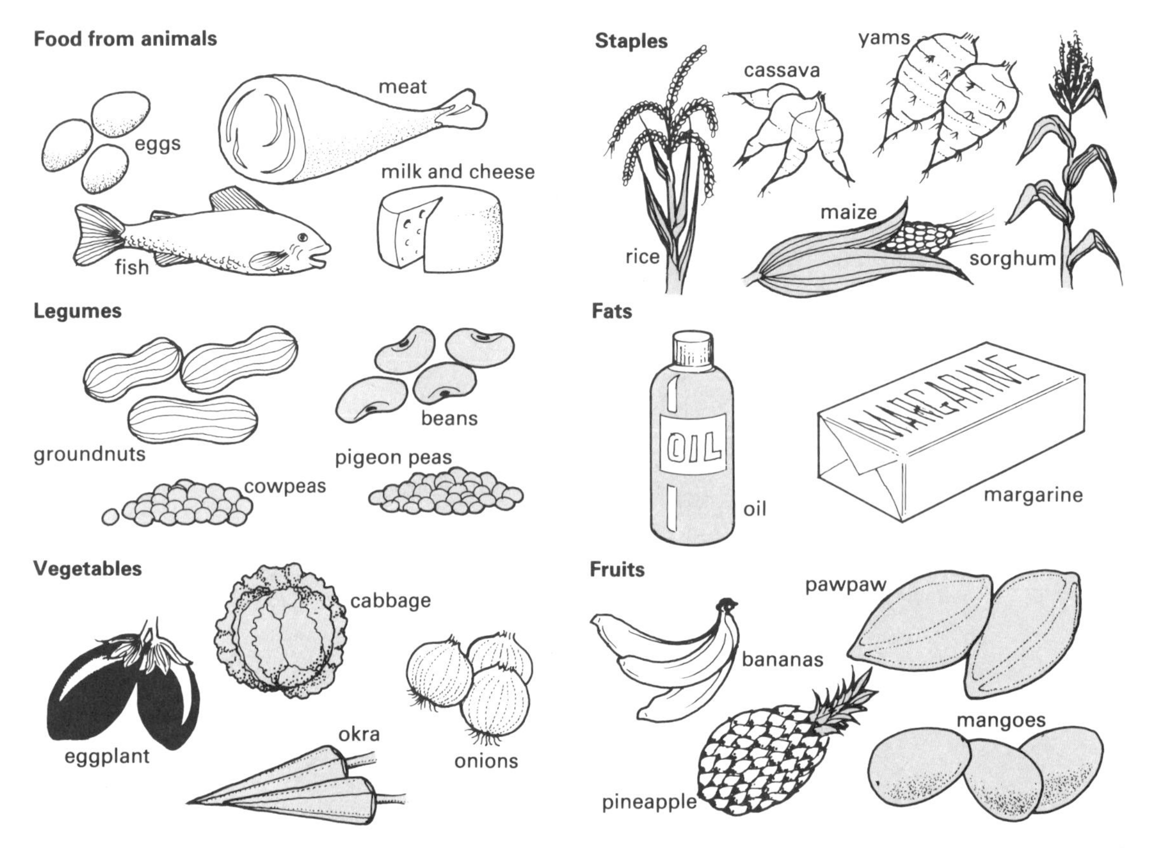

Figure 1.11 Some of the foods eaten in Sierra Leone. Foods from each group are needed for a balanced diet. All of these foods must be provided by agriculture

Perhaps you could draw up a list of the things you eat this week. Are you eating a balanced diet? Discuss this with your teacher.

In Sierra Leone the population is growing quickly and the number of people in the towns and cities is increasing. Our farmers must produce more food to keep everyone alive (Figure 1.12).

Figure 1.12 Everyone in our towns and cities must be fed by agriculture

Providing income and employment

Agriculture provides income and employment for over 70% of the population of Sierra Leone. Most of these people work independently as farmers, growing crops or raising livestock. Others are employed by organisations such as the Ministry of Agriculture or work on agricultural development projects such as the Integrated Agricultural Development Project (IADP); these people help farmers to increase production.

Some men and women may be employed indirectly in agriculture. Such people may be agents (buyers) who purchase cash crops (crops which are often exported) like coffee, cocoa, ginger and piassava or may buy agricultural goods for sale in local markets. Others use agricultural produce to make finished products which are sold; local shoe makers, for example, use the hides or skin of cattle to make bags, sandals and shoes. Do you know of any people who are indirectly employed in agriculture? What do they do?

Producing goods for export

Farmers in Sierra Leone often grow crops to sell for cash; many of these crops are grown specially for export. When Sierra Leone sells produce overseas it earns valuable foreign exchange; this is used to buy goods not made in this country. Examples of export cash crops include cocoa, coffee, ginger, palm kernel and piassava. These products are sold to overseas countries by the Sierra Leone Produce Marketing Board (SLPMB). The foreign exchange earned by the sale of these products may be used to buy things not made in Sierra Leone, such as cars, lorries, tractors, typewriters etc. Can you think of things you use every day which must be imported? Where do you think they come from?

We must also remember that if we can produce more food in Sierra Leone we need not import so much. In this way we can save our foreign exchange and spend it on other essential items. You can see that our farmers play a very important part in making our country wealthy.

Providing drugs and herbs

Certain plants are grown to provide us with useful medicines which are able to prevent illness and even cure a number of diseases. For example, ginger may be used to treat the common cold and sore throat. Some African farmers grow pyrethrum which is used to make a safe and natural insecticide.

Providing raw materials

Farmers in Sierra Leone supply many factories with the basic raw materials which are turned into finished products. Agriculture is therefore closely connected to some of our country's most important manufacturing industries. For example, the Chanrai Chemicals Factory uses palm oil, produced on our farms, to make soap; the Aureol Tobacco Company uses tobacco grown locally for the manufacture of cigarettes; and the Star Life Candle Factory uses animal fats to make candles. The Sierra Leone Produce Marketing Board Palm Kernel Oil Mill uses palm kernel to make maseray oil and the residue is used as a feedstuff for livestock.

In this section we should also remember the agricultural products we export to foreign countries; crops like cocoa and coffee will probably have to be processed overseas.

ACTIVITY 5

Copy Table 1.2 into your books. Complete the table as far as possible by naming an agricultural raw material, the factory or industry which uses it and the finished product or products. One example has been provided for you. To help you to start we have given you two finished products (beer and sugar). Were you able to complete the table on your own?

Table 1.2 Raw materials for manufacturing industries (Activity 5)

	Agricultural raw material	Factory/industry	Finished product(s)
1	Cacao	We Yone Cocoa Company	Cocoa powder
2			Beer
3			Sugar
4			
5			
6			
7			
8			

Providing feed for livestock

Our farmers produce a wide range of feedstuffs for the livestock in Sierra Leone. These feedstuffs may be grown by the farmers for feeding directly to animals or may be by-products from the processing of agricultural produce (Figure 1.13).

Like humans, livestock must eat a balanced diet and each type

Figure 1.13
Livestock feedstuffs:
(a) sheep and goats grazing on good pasture
(b) a farmer mixing pig feed

of animal has its own requirements; we will study these special needs in later books. Pigs may be fed with a wide variety of foods which might include rice bran, palm kernel cake, pumpkin and paw-paw. Groundnut cake and maize meal are suitable feeds for chickens. Cattle, sheep and goats eat grasses and legumes which may be specially grown by farmers or the animals may be allowed to graze in the bush. Our sea-fishing industry provides farmers with fish meal which is a valuable feed for animals; fish meal is produced by the drying and grinding of waste fish into a fine powder. It is an important source of protein.

Providing building materials

Increasingly, our farmers grow trees along with their normal agricultural crops; this is known as agroforestry and is an important development in our agriculture. These trees provide the farmers with building materials, fuel and animal feed and help with the fertility of the soil. We have seen that farmers use the land surrounding their villages to supply many of their needs. Can you name five types of material a farmer takes from trees and other plants to build a house?

Providing materials for making clothes

Some agricultural products are used to make clothes; for example, cotton is spun into thread, which is then woven into cloth for making clothes. You will probably have seen cotton being grown for this purpose. Have you ever watched people making country cloth? Is this a traditional activity in our villages? Animal skins from cattle and pigs can also be used for making clothes and shoes.

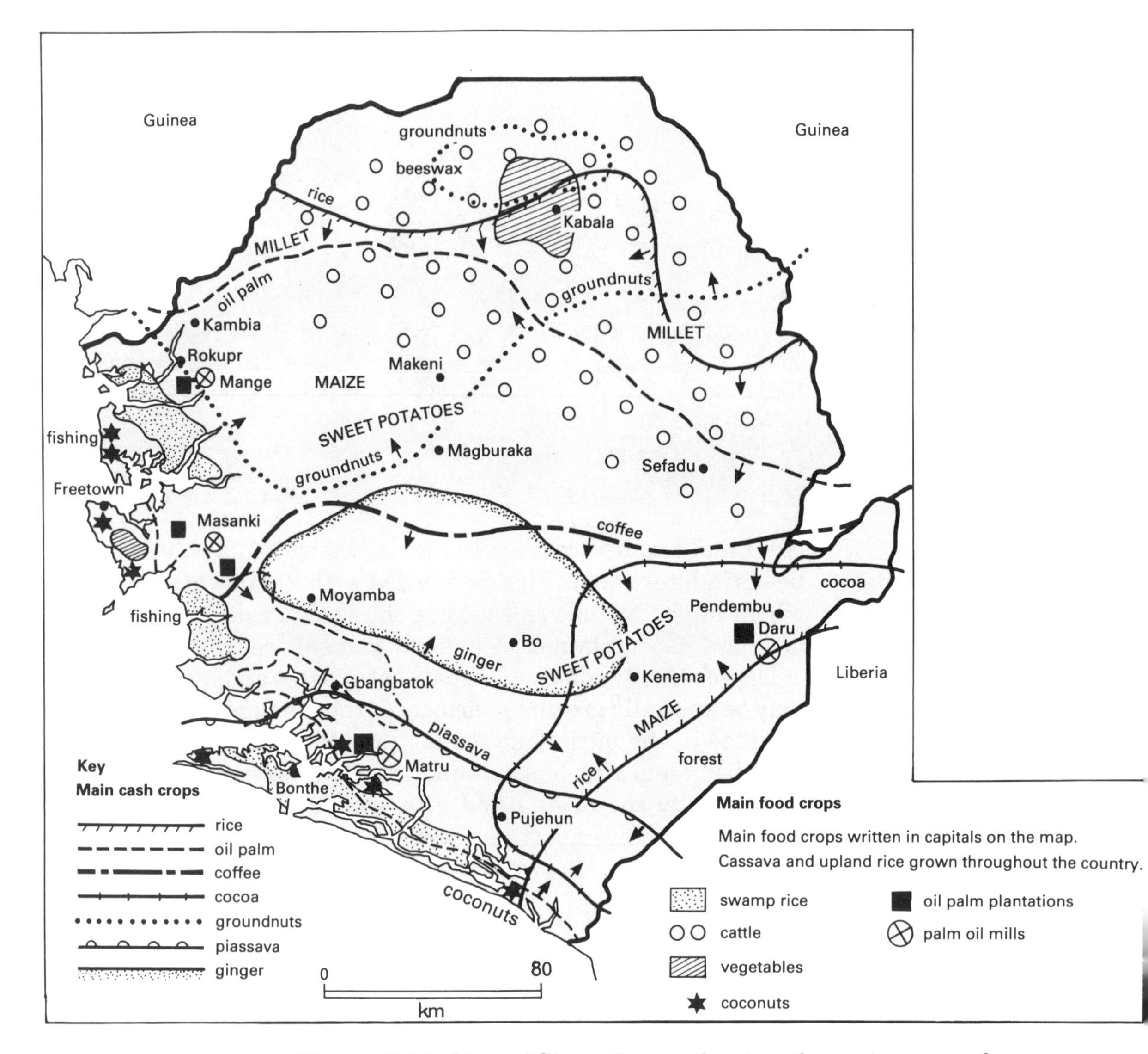

Figure 1.14 Map of Sierra Leone showing the main areas of agricultural production

Areas of agricultural production in Sierra Leone

Look at the map in Figure 1.14 showing the main areas of agricultural production in Sierra Leone.

Can you find where your own town or village is located on the map? Are the crops shown on the map grown in your area? Do you know of other crops grown in your area? Study the map carefully and try to form a picture in your mind of what you see. Activity 6 is based on this map.

ACTIVITY 6

Having studied the map carefully, answer the following questions. Do not look at the map until you have finished the whole activity!

Four possible answers are shown for each question, but only one is correct; select the correct answer and write it down in your exercise book.

1 In which area are coconuts grown?
  a the north
  b close to the border with Liberia
  c the coastal area
  d around the town of Bo

2 Around which town is the main vegetable growing area to be found?
  a Moyamba c Makeni
  b Pujehun d Kabala

3 How many oil palm plantations are shown on the map?
  a 5 c 7
  b 6 d 8

4 Where is the main cattle raising area?
  a north c east
  b south d west

5 In which region is coffee mainly grown?
  a northern region close to the border with Guinea
  b coastal region around Freetown
  c western region around the town of Rokupr
  d southern region of Sierra Leone

6 In which towns are there palm oil mills?
  a Moyamba, Magburaka, and Makeni
  b Masanki, Mange and Matru
  c Kenema, Kabala and Kambia
  d Bonthe, Bo and Sefadu

The global position of Sierra Leone

If we look at the map shown in Figure 1.15 we will see that Sierra Leone is to be found close to the equator on the western coastline of the African continent. The equator is an imaginary line drawn

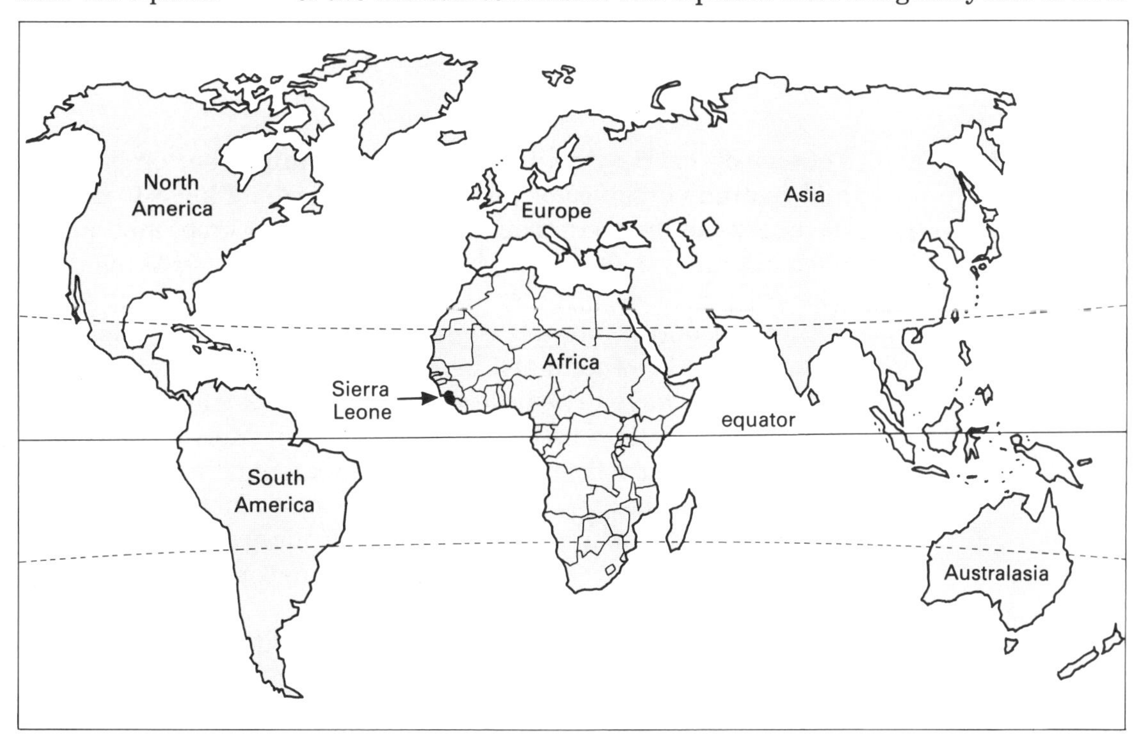

Figure 1.15 Map of the world showing the position of Sierra Leone near the equator

around the middle of the Earth which passes through all the places that are closest to the sun.

As Sierra Leone is so near the equator the climate (weather) is warm all the year round. The weather in our country is also affected by our closeness to the Atlantic Ocean, by the surrounding land and by the lowlands and hills within Sierra Leone. It is very important that we understand the climate in our country because it has a great effect on our agriculture and determines what farmers do (planting, growing and harvesting). You can see now why farmers are always talking about the weather!

Much of Sierra Leone, especially in the south, is covered with lowland rainforest. The north is more open, with fewer trees. It is a little drier and the land is at a higher altitude. Rainfall, which is so essential for agriculture, is spread over six months of the year, and encourages tree and crop growth. In the previous section we saw that the southern and eastern parts of Sierra Leone, with adequate rainfall, grow crops such as coffee and cocoa. Rice grows well in the swamps and wet lands of the coastal parts of the country. In the north, where it is drier, millet is grown and cattle rearing is of major importance. What is the climate in your area? Perhaps your class can record the weather during the school year.

The type of agriculture carried out in an area is not only determined by the climate; the type of soils found in the area and the shape of the land (flat or hilly) will also play a part in determining what can be grown. In the next chapter we will learn how soils are formed and we will see that the weather also plays an important part in this process. In the next chapter we will also look at the soils in your area.

Farmers and the environment

Let us look back at the diagram of the farming year on page 10, and remind ourselves of the work that farmers have to do. We have seen how farmers spend their year growing crops and raising livestock and we have thought about the many other things that people have to do in order to live and for which they use the natural resources around them. For example, if food is to be cooked, then firewood must be found; if buildings are to be put up then poles have to be collected from the forest; and if washing is to be done, water must be collected from a river or well. Farming communities must, therefore, rely on the surrounding countryside to supply their daily needs. If we are going to increase our agricultural production and meet the needs of our growing population it is important that we make sure these natural resources will continue to supply all our needs.

There is increasing evidence that the natural environment in Sierra Leone is being destroyed by too much exploitation (overuse). Our forests are being cut down to supply wood for fuel

(a)

(b)

(c)

Figure 1.16 Different types of land in Africa:
(a) rainforest in Sierra Leone
(b) good agricultural land in Sierra Leone
(c) a denuded and eroded area after trees have been cut down, in the Sudan

and timber for building, too many wild animals are being killed for food and the rains are washing away a lot of valuable soil from our farmers' fields. This cannot be allowed to continue; we must learn to take no more than we need, to harvest only what the countryside can replace and to conserve (save and look after) all our natural resources. Only in this way will future generations be able to live comfortably. If we ignore these rules there will be less and less for everybody until, eventually, we will have created a desert in which nobody can live; this has already happened in many parts of Africa. Conservation, taking care to preserve our national resources so that there will be plenty for future generations, is a subject you must study very seriously (Figure 1.16).

Ask some of the old people in your village to tell you what life was like when they were young. Were there plenty of wild animals and fish? Ask them what the countryside was like in those days. What changes have they seen? What has caused these changes? Are these changes for the better or worse?

Here are a few questions for you to answer. Fill in the blanks in the sentences using the correct words from the box:

desert **conservation** **erosion** **firewood** **exploitation**

1 ...... is caused by rain washing away the soil.
2 Without ...... you cannot boil your rice.
3 If we cut down trees, use our soils and take wild animals from the forests it is called ......
4 ...... means looking after our natural resources.
5 A ..... is created when all vegetation and trees have been removed from the land.

What can you do to help?

It is not easy for a young person to make a contribution to the conservation of his or her country. The problems are very great and often must be tackled at the highest levels of government or by the leaders in a community. However, you will be the adults of tomorrow and it is important that you understand the issues and discuss them with your parents and community leaders.

Even though it may be difficult, there are many things you can do to help conservation. Here are a few ideas:

- you can plant trees to replace those cut down
- you should not allow bush fires to burn more vegetation than is needed for the growing of your crops
- you can learn how to become a good farmer and how to get more out of your land
- you should learn to respect all wild creatures and plants
- you should not kill wild animals unnecessarily and you should not harm those that are rare
- you should not pollute the environment by throwing away litter

How do you think these ideas help the environment? Can you think of any other things you can do to help conservation? Do you think your school class can carry out a conservation project?

Now close your book and see how many conservation ideas you can remember. Get a friend to test you.

Opportunities in agriculture

We have seen that agriculture is the most important industry in Sierra Leone and that it employs the largest number of people in our country. There are, therefore, many job opportunities in this industry.

Sierra Leone has a great need for practical farmers who have a knowledge about modern agriculture; this is why it is so important to study farming at school. While there is a great deal you can learn from watching farmers and helping them at work, you must also learn about the new ways of producing crops and livestock; only in this way will you become a successful and wealthy farmer able to make a contribution to Sierra Leone's development and have a secure future.

If you work very hard during your time at Junior Secondary School a number of other job opportunities may arise. When you leave you could apply to study at the Certificate Training Centre at Njala (where you study for a certificate in agriculture) or at the Njala University College (where you would study for a degree in agriculture). At these places you would have the chance to study a wide range of agricultural subjects which would include machinery, livestock, crops, pest and disease control, veterinary science and farm management.

(a)

(b)

Figure 1.17 Agricultural specialists:
(a) extension agent
(b) research worker at a cowpea research station

When you finish your course you could choose to join the Ministry of Agriculture and become a technician and, as an extension officer, give guidance to farmers. Alternatively, you might choose to join the Ministry of Education and teach agriculture in secondary schools, the Teachers Training College, the Certificate Training Centre or Njala University College.

With further study you could become a specialist (Figure 1.17). You might, for example, work in the Veterinary Service to treat sick animals and prevent disease.

Some key words and terms in this chapter

Agriculture The cultivation of the land, and the raising of crops and livestock.
Market A place where people gather to buy and sell goods.
Farming year The annual cycle of activities followed by farmers.
Transplanting The removal of a young plant from a nursery bed and planting it out into the field.
Irrigation The supply of water to crops through artificial canals, ditches or pipes.
Balanced diet The correct mixture of food and drink that provides animals with all their nutritional needs.
Conservation The care and preservation of our natural resources.
Environment The surroundings in which we live and work.
Cash crops Crops that are grown specially for export to earn foreign exchange
Foreign exchange Money earned for a country by the sale of exports to another country.
Equator An imaginary line drawn around the earth joining all points that are nearest to the sun.
Climate The weather conditions that affect an area.

Exercises

Puzzles

1 Copy the puzzle grid into your exercise book and try to complete it.

Across
1 Their skins are made into leather (6)
2 Their meat is called pork (4)
Down
3 They are kept in small cages or hutches (7)

2 Copy the puzzle grid into your exercise book and try to complete it.

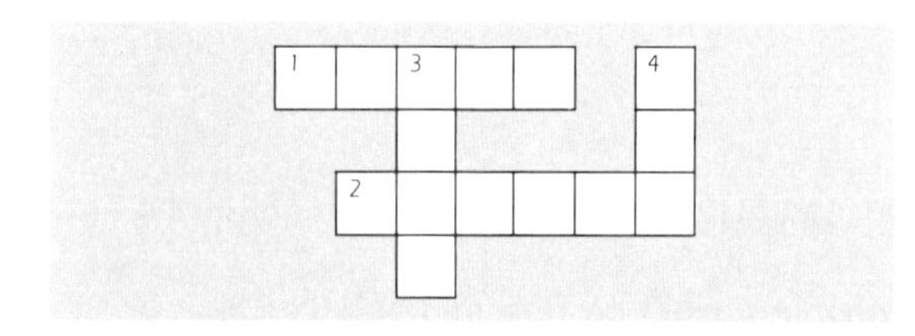

Across
1 In which part of the country is open land most often found? (5)
2 One of the grain crops grown for human food (6)
Down
3 Falls for six months in Sierra Leone (4)
4 A small wild animal which can damage crops in storage (3)

3 Try to complete this puzzle by writing the missing words into your exercise book.
a R. . . is one of our main foods.
b A young goat is called a k. .
c A c. . gives us milk
d Your shirt is often made of c. . . . .

Multiple choice questions

Write the correct answer in your exercise book.

4 Which of the following crops gives us oil?
a sorghum
b rice
c groundnuts
d millet

5 Which of these animals does **not** eat grass as the main part of its diet?
a cows
b pigs
c sheep
d goats

6 Which of the following definitions best describes conservation?
a cutting down trees
b making cloth from cotton
c taking care of our natural resources
d throwing away food we do not want

7 Which crop grows best in a swamp?
a oil palm
b coffee
c maize
d rice

8 Which of these is a wild animal?
a sheep
b leopard
c bull
d chicken

9 Which of the following is **not** a job opportunity in agriculture?
a The rearing of livestock.
b The cultivation of crops and crop research.
c The production of finished products from agricultural raw materials.
d Spreading information about modern agricultural techniques.

10 Which of the following was **not** true of the very earliest stage of agricultural development?
a Early people gathered fruits for consumption.
b Early people hunted animals for food.
c Early people rubbed stones together to produce fire for cooking.
d Early people kept livestock in specially prepared pens.

Matching items

Write the answers in your exercise book.

11 Complete the following statements by selecting the most appropriate word or words from below.

a Agriculture provides ...... for industry.

b Sierra Leone earns valuable ...... from agricultural exports.

c Agriculture provides ...... for the increasing population.

d Agriculture provides ...... for many people in Sierra Leone.

income and employment
foreign exchange
raw materials
food and clothing

12 Match the statements with the most appropriate answer given below.

a An important agricultural export crop in Sierra Leone.

b What is agriculture?

c The occupations of early people.

d An important raw material for the making of candles.

e The name of the most important cereal crop in Sierra Leone.

f Used by local shoe manufacturing industries.

g One way in which people can be engaged in agriculture.

h One of the roles of agriculture in the economy.

(i) animal fat
(ii) rice
(iii) hides and skins of livestock
(iv) teaching agriculture in schools
(v) provision of feed for livestock
(vi) gathering of fruits and hunting
(vii) cultivation of crops, rearing of livestock and maintenance of soil fertility
(viii) palm kernel

Points for discussion

13 What natural resources are exploited by your local community? Who controls the use of these resources and how are they prevented from overuse?

14 What do farming people do all day? What is the normal pattern of their daily activity?

15 Do you know of any traditional dances, stories or songs that tell about agriculture? Do people sing about agriculture when they are working in the fields?

CHAPTER 2 The origin and nature of soils

Introduction

In this chapter we will study the soil. We will discover that soils are the basis for farming, for without them we could grow no crops, animals would have nothing to eat and we would all starve! We will find out how soils are formed, what soils are made of and what properties they have. We hope you will enjoy doing some tests on soils from your district.

The formation and origin of soils

Figure 2.1 A muddy river carrying soil particles

Millions of years ago the surface of the earth was covered with hard rock. When exposed to the forces of nature the surface gradually broke down into small pieces called particles. In time, these particles of rock (made up of inorganic or mineral matter) formed soil. Let's find out how this happened!

The process took thousands of years to happen and was caused by the effects of water, wind, temperature and living organisms on the surface of rocks. The process is called weathering because, as you can see, it was mainly caused by the weather! Where you live, you may be able to see rocks sticking above the soil (look especially on the tops of bare hills); these rocks are slowly being weathered and, over a very long period of time, they will break down to form soil. The type of soil that is formed will depend on the type of parent rock. Look at the surface of some exposed rock. Can you see any signs of weathering?

The rocks in your district may well have produced the soil that lies around them. In other areas, the soil may have been carried many hundreds of kilometres away from its parent rock. Water and wind may have played a part in this; soils that have been moved in this way are called transported soils.

You will have seen small streams of muddy water flowing after heavy rains. If you collect some of this water in a bottle and let it stand for a time you will see that it will gradually clear as the mud falls to the bottom. The mud is formed from minute particles of soil. Winds may also pick up soil particles and as the winds die down the particles fall to the ground. In both cases the particles may be put down some distance from where they were formed. These are examples of how soils are transported (Figure 2.1).

In fact, most of the soils in Sierra Leone have not been transported but have been formed from the rock that lies directly below the surface.

The weathering of rocks

Let us remind ourselves of the four things that cause weathering (Figure 2.2); we will then look at each one in more detail:

- temperature changes: heat and cold
- water: rain, moving water, and freezing water
- wind
- living organisms: plants and animals

Figure 2.2
The main processes involved in soil formation

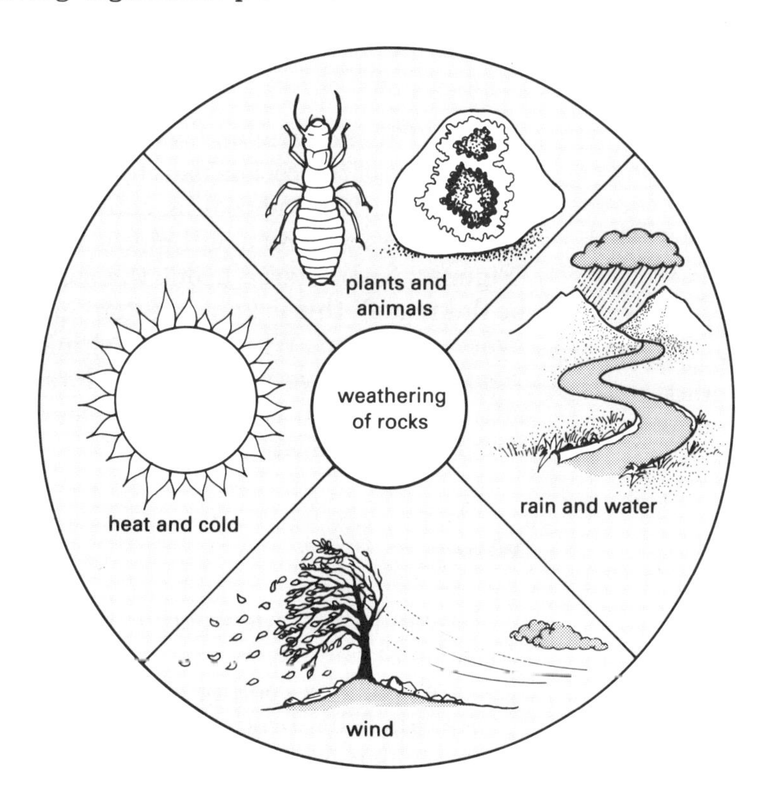

Temperature changes

When rocks are heated up and cooled down they expand (grow bigger) and contract (grow smaller) by a very small amount. This happens during the day when the sun shines and the rocks grow hot and during the night when the rocks cool down again. The rocks may also be cooled down quite quickly if there is a shower

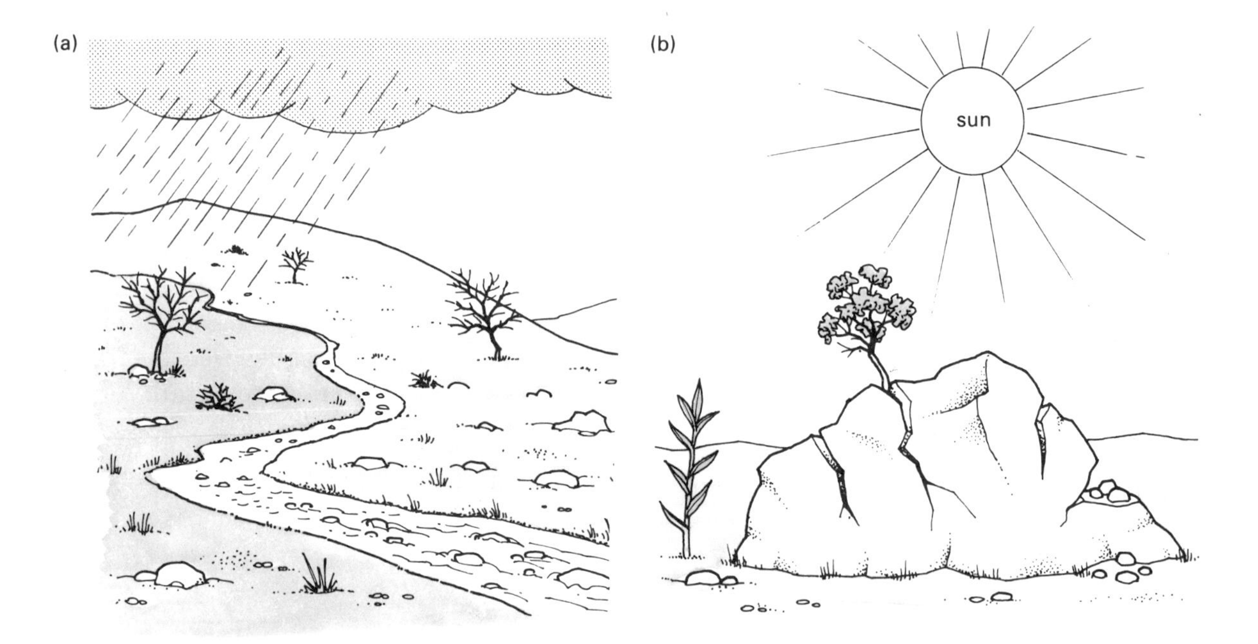

Figure 2.3
The weathering of rocks:
(a) running water wears rocks away
(b) heat from the sun causes rocks to crack and break up

of rain. This heating and cooling is never equally balanced and the surface of the rock is usually hotter or colder than the middle. All this heating and cooling causes tiny cracks to appear on the surface of rocks and small pieces fall to the ground to form soil (Figure 2.3). In Sierra Leone the surface of rocks may be seen to be flaking off; this is known as 'onion skin' weathering. Have you ever seen this? Why do you think it is called 'onion skin' weathering? Have you ever felt how hot a rock can get if it is sitting in the sun? Have you ever seen a rock steaming as it cools down after a shower of rain?

Water

Water can help in the weathering of rocks in a number of ways. We have already seen how rainwater can help in the breaking down of rocks by cooling.

As rain falls through the air it takes in some carbon dioxide (a gas) which forms a weak acid (a liquid which can eat away things it comes into contact with). Another weak acid is formed during thunderstorms. These acids attack the surface of rocks causing small particles to break off.

When it rains, water collects to form streams and rivers. Over time this moving water can break off particles from the surface of rocks which add to the soil. Have you ever seen smooth round rocks in the bed of a stream or river? What has caused this? Have you ever seen a river flowing through a gorge between rocks? How has this happened?

In countries that are much colder than Sierra Leone water can play another important part in the weathering of rocks. First, however, we must understand that when water freezes and turns to ice it expands; if the water is trapped in a space it can exert a great force as it freezes. Have you ever seen ice? Look in the ice box in a refrigerator. If you were to try and freeze a bottle of water in a deep freezer there is a danger that the bottle may break. Why is this?

In cold countries water may be trapped in the cracks in the rocks; as the water freezes it expands causing the rocks to break apart. When the ice melts, the rock particles fall away to make soil.

Wind

Strong winds can pick up soil particles and carry them along. The sharp pieces you get in your eyes during a dust storm are these tiny rock particles. Rock particles may hit against larger rocks, slowly wearing them away. In some deserts in Africa there are good examples of this kind of weathering (Figure 2.4).

Figure 2.4
Rocks eroded by dust and wind in the Sudan

Living organisms

Both plants and animals play a part in the weathering of rocks and in the formation of soil.

Plants

Very simple plants, called lichens, grow on the surface of rocks.

Figure 2.5
Plants that help to break down rocks

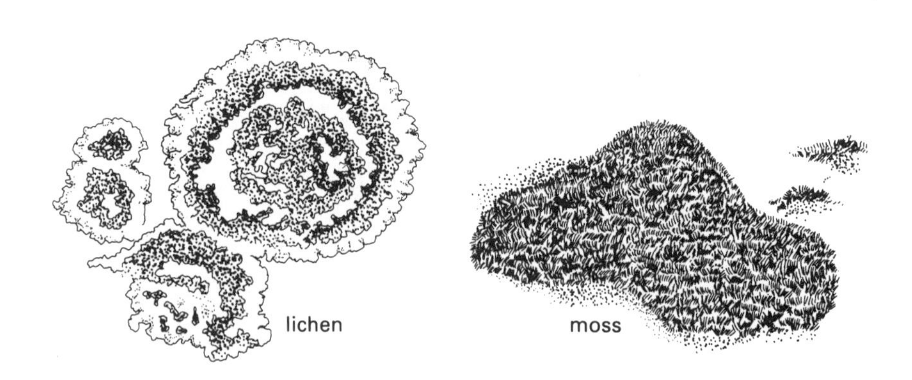

These simple plants have no roots but obtain the food they need by dissolving (eating away) the surface of the rocks; over time this causes weathering (Figure 2.5).

Other plants, growing on rocks, take in the water and minerals they need through their roots. These roots creep down into the cracks in the rocks and make them wider and cause rock particles to break off. Plant roots also contain chemicals capable of dissolving minerals, which adds to the process of weathering.

Animals

All soils contain many living organisms which are always moving through the soils. The action of animals, such as worms, termites, millipedes, beetles and rats, adds to the breaking down of soil into finer and finer pieces (Figure 2.6).

Figure 2.6
Animals that help to break down soil into finer particles

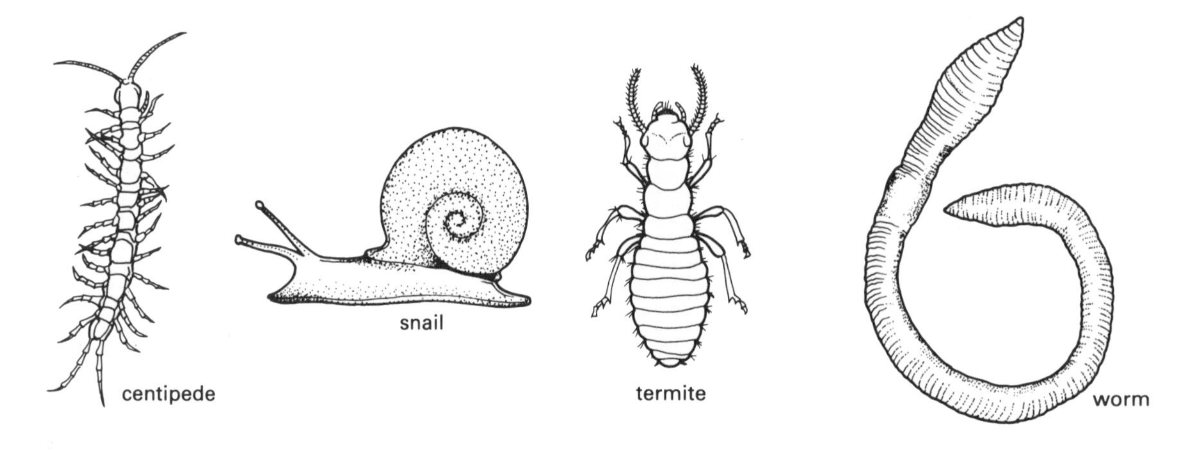

Mechanical, chemical and biological weathering

The action of wind, rain, heat and cold on rocks is known as mechanical (or physical) weathering. The action of acids from the air is known as chemical weathering. The action of plants and animals in breaking down the soil into finer particles is called biological weathering (Figure 2.7).

Figure 2.7
Diagram to show how mechanical, biological and chemical weathering processes overlap

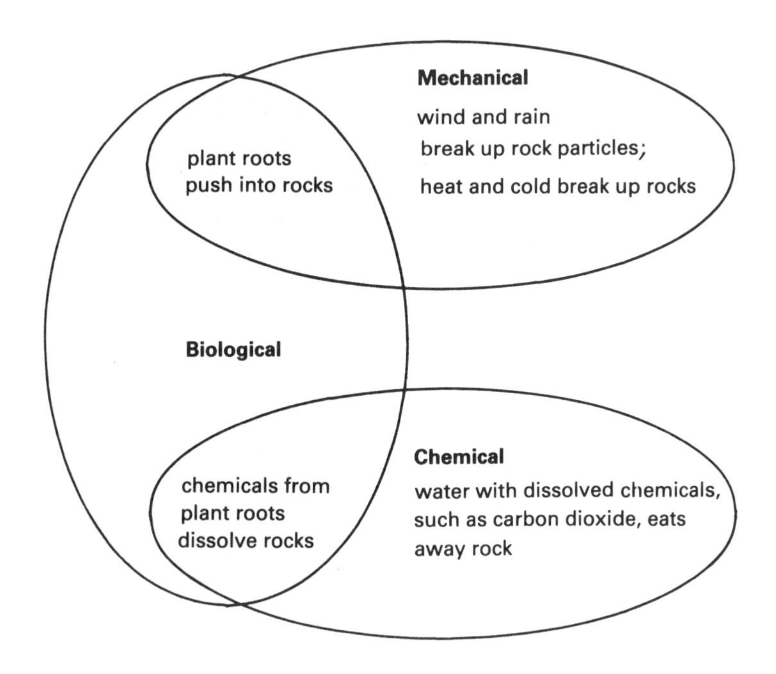

Definition and components of soil

What is soil?

If you were asked this question, what answer would you give? You might be tempted to say "the ground" or "something you find on a farm in which plants grow". There is not much wrong with either of these definitions but as agriculturalists or farmers we need to be more exact. We could say, perhaps, that soil is "the natural formation covering most of the land surface of the earth and which supports plant growth" or we could say that a soil is "the material in which plants live and which provides them with support, water and plant food".

Components of soil

We have seen that soils are made up of organic (from living things) and inorganic (from non-living) materials. Organic material includes worms, termites, ants, bacteria, fungi, algae, and other small forms of life, as well as all the material from dead plants and animals, especially dead leaves and other plant remains. The inorganic material consists of pieces of weathered rock of different sizes, and water. Have you ever looked really closely at some soil? Go and dig some up and look very carefully at it. What do you see? What does it feel like? What colour is it?

As agriculturalists we are interested in soils and we need to understand how they are made up. You will have noticed that a soil is made up of many different sized pieces. You will probably also have noticed that soils from different places are quite distinct;

some soils will be stony and some soils will be quite sandy. How would you describe the soil from your region?

To help us describe soils, scientists have given special sizes to each of the soil particles. Clay particles are the smallest, followed by silt, then fine sand, coarse sand and finally the largest group is made up of stones and gravel. These sizes are shown in Table 2.1.

Table 2.1 Sizes of soil particles – the International System

Particle	Size
Stones and gravel	2 mm + in diameter
Coarse sand	2 mm to 0.2 mm
Fine sand	0.2 mm to 0.02 mm
Silt	0.02 mm to 0.002 mm
Clay	less than 0.002 mm

These figures may be useful to you if you study agriculture at a college but at this stage they are only given to you for interest. Do not try to learn them by heart!

ACTIVITY 1

Near the left hand side of your notebook draw a vertical line measuring 20 cm; this represents the size of a coarse sand particle. Now a little distance away draw another line 2 cm high; this represents the size of a particle of fine sand. Beside the 2 cm line draw a very small line only 2 mm high; this represents a particle of silt. Finally, with a sharp pencil point make a tiny dot beside the 2 mm line; this is the size of a clay particle. Label your lines. You can see from the lines and the dot the relative sizes of the different particles. If you can only just see a coarse sand particle in a sample of soil you can imagine how small a clay particle is!

Types of soil in Sierra Leone

There are three main types of soil in Sierra Leone:

- sandy soils
- clayey soils
- loamy soils

Sandy soils

These soils are mainly made up of the larger soil particles (stones and gravel, coarse sand and fine sand) which you can see with your naked eye. The soil particles have large gaps between them which allow water and air to move freely in the soil. Sandy soils allow water to drain away quickly and do not hold much moisture; sandy soils do not support crops very well during the dry season or when there is a drought.

Figure 2.8
Sandy soil with large particles and coarse texture (left)
Figure 2.9
Clayey soil with very small particles and fine texture (right)

Sandy soils are often very poor in plant food and are not usually good for crop production. They can be improved by adding manures (compost or farmyard manure). These materials fill in the spaces between the soil particles, help to hold moisture and provide plant food as they rot down. Sandy soils are mostly found in upland areas of Sierra Leone.

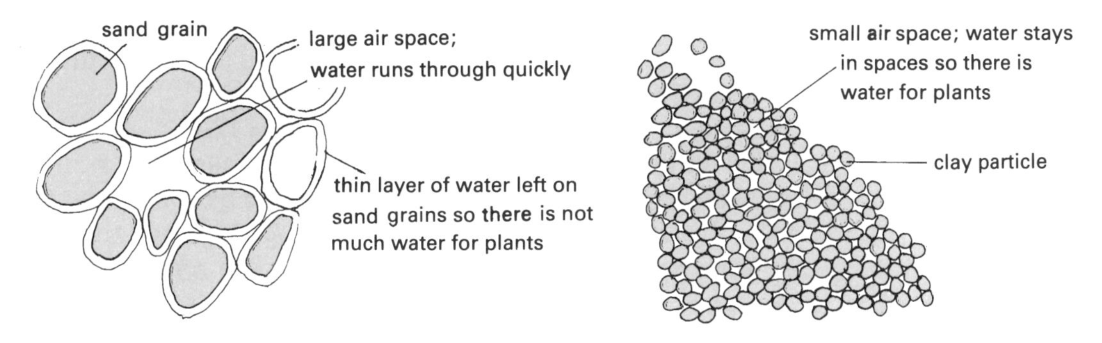

Clayey soils

These soils are made up of a large amount of the smallest soil particles (clay). These small particles pack very closely together leaving only tiny spaces between them. Clayey soils have very different properties to the sandy soils we have just read about.

Clayey soils are sticky when wet and hard when dry and do not allow water and air to pass freely through them (Figure 2.9). Clayey soils are therefore difficult and heavy to work with and may quickly become waterlogged (soaked with water). Clayey soils, however, can be very fertile (contain lots of plant food) and can hold a lot of water. This means that they can grow very good crops, even during the dry season; swamp rice and sugar cane grow very well in these soils. Clayey soils are mostly found in the swampy areas of Sierra Leone.

Loamy soils

These soils are made up of an equal mixture of sand, silt and clay particles. They have all the good points of sandy soils and clayey soils without any of the problems!

Loamy soils have enough space between the soil particles and while they can hold water and air, both can move freely through the soil. We can say that loamy soils have a good soil texture (the mixture of soil particles; the balance between sand silt and clay) and a good soil structure (the arrangement of particles in a soil). Loamy soils should be dark in colour, showing that they have a lot of organic matter; this makes them rich in plant foods.

Loamy soils, being easy to work, fertile and able to hold water and air, are considered the best soils for crop production. Loamy soils are found in upland areas and in swamps in Sierra Leone.

Let us now find a sample of soil and carry out a test to see how much of it is sand, silt, or clay.

ACTIVITY 2

1 Find a container like a measuring cylinder or a clear plastic bottle. Your teacher may help you with this.
2 Collect some fresh soil and put it into the container until it is ⅓ to ½ full. Do not use dry soil.
3 Add clean water until the container is ¾ full.
4 Thoroughly mix the contents of the container by shaking or stirring. The soil and water may be mixed by stirring with a rod or by shaking the container (you may put a cap on the bottle or place your hand over the neck of it). The contents should be mixed for several minutes.
5 Leave the container on a flat table for 1½ hours.
6 Draw in your notebook what you see in the container and then compare your drawing with Figure 2.10.

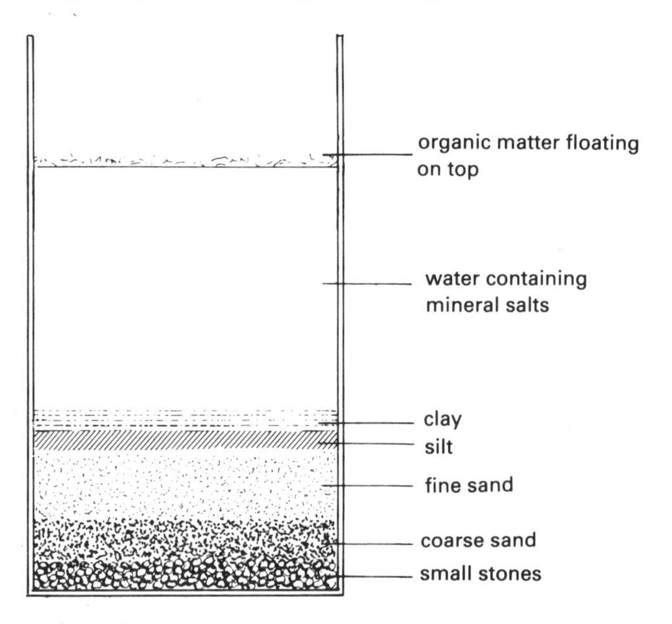

Figure 2.10
Does your soil separate out like this?

You will see that most of the particles of soil have sunk to the bottom. If you look very carefully you will see that they have formed layers; the heaviest and largest lie on the very bottom while smaller particles rest on top of them. At the bottom lie the stones and gravel with the coarse sand lying between them. The next layer should be the fine sand and above this lies the silt. The clay is the last layer to settle out and should be lying on top.

If the water above the settled soil is clear, then all the mineral matter has settled; if it remains cloudy, then many of the clay particles will still be in suspension (trapped in the liquid); the clay particles may take many hours to settle out completely. Floating on top of the liquid you may see a thin line of dark particles; this is the organic matter from the soil.

How did the results from your test compare with the diagram? Did it look the same? You must remember that no two soils are the same and we would normally expect different results.

Take a ruler and measure the thickness of each layer and record them in your exercise book. Do you think your soil was a sandy soil, a clay or a loam.

ACTIVITY 3 Looking at different soils

For this activity you will need dry samples of sandy, silty and clayey soils. Your teacher will give you these. You will also need some water to mix with the soils. Make sure that you do not mix up the three samples; label each one carefully. Carry out the following simple tests on each soil sample.

Sandy soil

1 Look at the sample thoroughly; you should be able to see lots of tiny grains of sand.
2 Rub a little between your fingers. It will fall through quite easily. You will find that a sandy soil has a coarse texture. Now wet a small sample and rub it between your fingers; it will feel gritty.
3 Take another portion in your hands and try to roll it into a ball. You will see that this cannot be done and the soil falls apart.

Write down in your notebook what you have found out about sandy soils.

Silty soil

1 Look very carefully at this soil sample. How does it differ from the sandy soil? Are the soil particles larger or smaller?
2 Place a little of the silty soil in your hand and pour on a small amount of water. Rub the surface of the soil with the tip of your finger. You will notice that it has a slippery or soapy feel. Why is this? If there is any sand in your sample, it may feel gritty, but if there is some clay in it there may be a sticky feeling.
3 Roll the sample between your hands and try to make a ball; this should be quite easy. Now try to make a roll or a stick. Next bend it into a ring; it will probably break.

Write what you have discovered about silty soils in your notebook.

Clayey soil

1 Look very carefully at this soil sample and compare it with the sandy and silty soils.
2 Moisten the clay sample and roll it between your hands. It will feel sticky, and will cling to your hands and colour them. The clayey soil has a fine texture. Perhaps you can now understand why clayey soils are so difficult for farmers to work with!
3 Having made a roll or stick from the soil sample, bend the roll into a ring. Does it break? No, it does not! Why is this? Clayey soils can be identified using this method.

Write down what you have discovered about clayey soils in your notebook. Each of the soils was really quite different.

Summary

A clayey soil will have more clay than sand and silt.
A sandy soil will have more sand than clay and silt.
A silty soil will have more silt than clay and sand.
A mixture of all three (sand, silt and clay) is the best type of soil for agriculture and is known as a loam.

ACTIVITY 4

Your teacher will give you a sample of a loam, which, as we have discovered, has a roughly equal mixture of sand, silt and clay.

If the soil sample is dry, moisten it with a little water. Repeat the tests you have done on the other soil types. How did the results compare? Describe in your notebook what you found.

At first Figure 2.11 may seem very difficult; this is not the case! It is really quite simple! The triangle is useful for identifying different soil types and giving them an accurate name. You will see that each side of the triangle represents one of the parts that

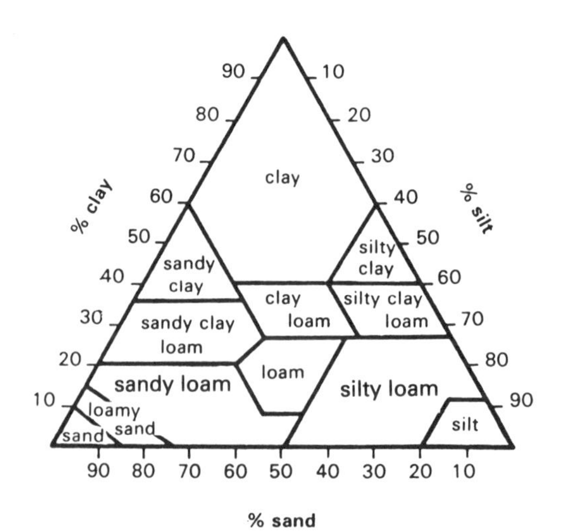

Figure 2.11
Soil classification according to texture

make up a soil: sand, silt or clay. Each side ranges from 0% to 100% depending on the amount of each in the soil sample.

In each corner you can see what a soil is called if it is made up of nearly all of one part (a sand, a clay and a silt loam). The rest of the triangle shows us what soils are called if they have different mixtures of particles. We can see that a loam appears in the middle of the triangle as it contains roughly equal amounts of each. If a clayey soil has some sand and silt in it we call it a 'clay loam'. If a soil is mostly silt with a lot of sand but a little clay it is called a 'sandy silt loam'. Get your teacher to test you with a few examples to see that you understand the diagram.

Draw or copy the diagram into your notebook. You will find it useful when you visit farms and try to identify the soils there.

As you travel through the countryside, look out for different soil types. On the seashore or at the sides of a river you might find a pure sand. If you go to a river swamp or an estuary by the sea you may find a pure silt. At a spring, where a stream runs out from the soil, you may find a clay.

The ability of soils to hold water

As young farmers we must know how water acts in different soils. This is essential if we are to know how to grow good crops!

We already know that some soils allow water to pass through them easily and that they hold little water for growing crops. We have seen that other soils will hold quite a lot of water but will not allow it to pass through easily. Let us carry out some tests!

ACTIVITY 5

The following activity is more easily carried out as a class activity or if you work in a few large groups (Figure 2.12).

Your teacher will give you three cups containing:

- dry fine clay
- dry sand
- dry loam

You will also be given:

- three identical clear plastic bottles
- some cloth
- some string

1 Before you can begin the test you should copy Table 2.2 into your notebook.

Table 2.2 Results of soil and water test (Activity 5)

Soil type	Water added (cm)	Water in base (cm) at given intervals (minutes) 5	10	15	20	25	30	35	40	45
Sand										
Clay										
Loam										

2 Take each bottle and tie a piece of cotton cloth over the open neck end. Now cut each bottle in half across the middle. You now have one half with a solid base and the other half can be turned over so that the neck end is at the bottom.
3 Fill the neck end of each bottle with the **same amount** of each sample of soil; the cloth will stop any soil from falling through. Label each bottle correctly.
4 Half fill the bottom sections of each of the bottles with the **same amount** of water, and make a mark on the sides to show the level of the water.
5 When you are ready, tip the water from the containers on to the three soil samples; do it carefully so that none is spilt. Immediately place the empty containers underneath the cloth covered ends of the bottles to collect the water as it runs through.
6 After the given periods of time record the depth of water that has collected in the bases.
7 When no more water flows through the samples you can complete the table in your exercise book. Write up how you carried out the test in your exercise book.

Figure 2.12
Experiment to find out how fast water passes through soil

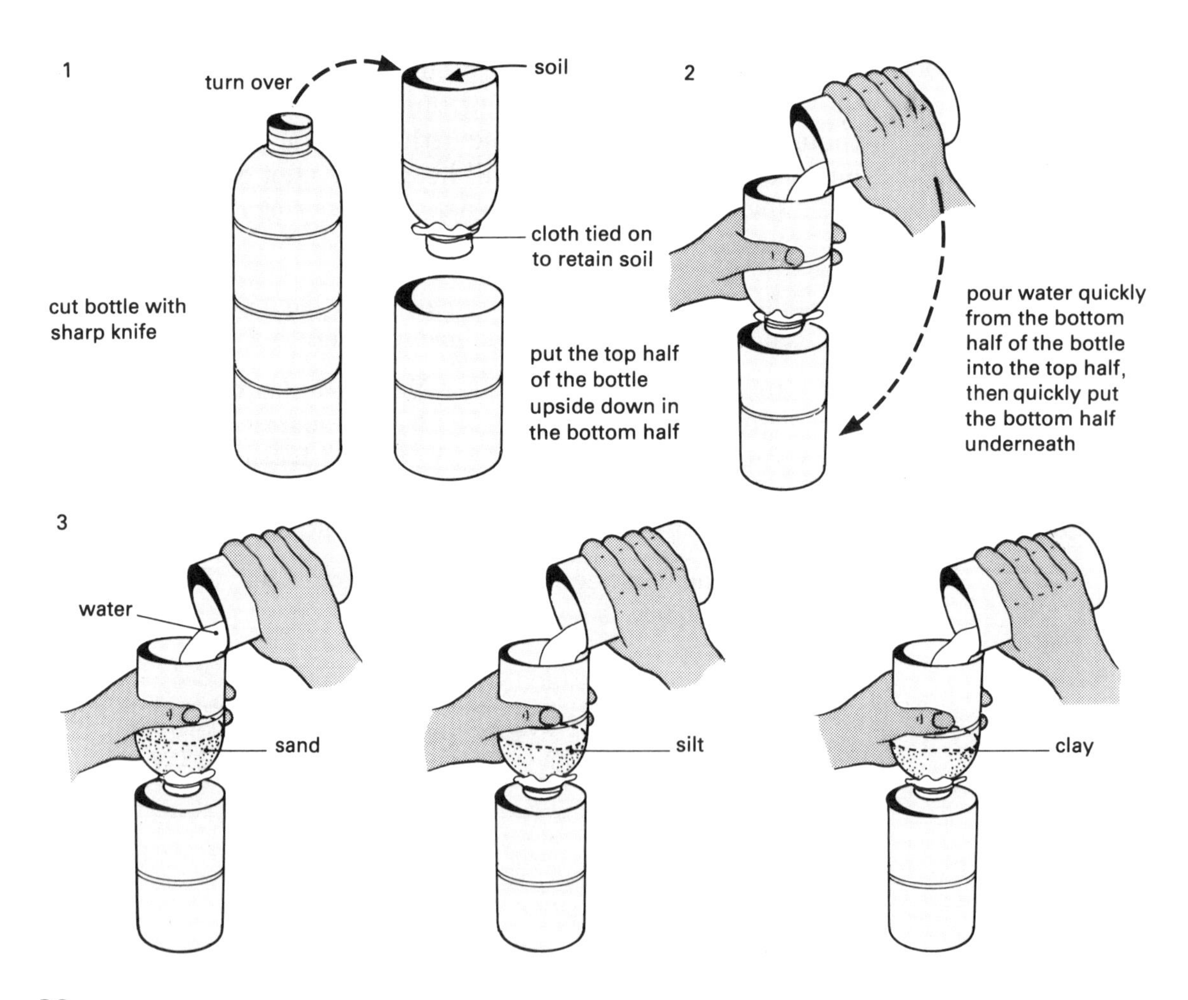

From the information you have written down you will be able to answer a number of questions:

- Which soil held most water?
- Which soil let almost all the water pass through?
- Which soil allowed water to pass through quickest?
- Which soil allowed water to pass through slowest?
- What can we tell about the soils by the different speeds at which water ran through?

What does this test tell us about these soils? Remember plants need water from the soil but do not like their roots to be soaked; plant roots need air as well. Which of the three soils do you think will be most suitable for plants? Which of the soils seems best for farming? Why is this?

We have just learnt that plant roots need air as well as water, but have probably never thought that the soil contains much air! However, if we think about it, there must be small spaces between the soil particles. These spaces, which soil scientists call pore spaces, are filled with air or water. Farmers need to keep a balance between the amount of water and air in a soil. How can they do this? How do farmers increase the amount of pore space in a soil?

ACTIVITY 6

How much air is there in soil?

This is a class or group activity. Your teacher will provide you with:

- a clean straight-sided tin
- a hammer and a nail
- a measuring cylinder
- sticky tape
- some water
- a sample of soil

1 Copy down the following table into your exercise book:

Table 2.3 How much air is there in soil? (Activity 6)

Top mark of measuring cylinder	(a)		____ ml
Mark of measuring cylinder after filling empty tin	(b)		____ ml
Volume of tin (or soil)	(c)	(subtract)	____ ml
Top mark of measuring cylinder	(d)		____ ml
Mark of measuring cylinder after filling soil-filled tin	(e)		____ ml
Volume of air in soil	(f)	(subtract)	____ ml

The volume of air in the soil can now be stated as a percentage:

Volume of air (f) ____ ml divided by volume of soil (c) ____ ml multiplied by 100 gives us the % pore space.

2 Fill the measuring cylinder to the top mark and record how much water it holds (a).
3 Carefully pour the water into the tin until it is completely full. Record the new level of the water in the cylinder (b).
4 Subtract these two figures to give you the volume of the tin (c).
5 Empty the tin. Using the nail and the hammer, make two or three holes in the bottom of the tin. This will let air out of the tin during the next operation.
6 Find a piece of ground that has recently been cultivated and

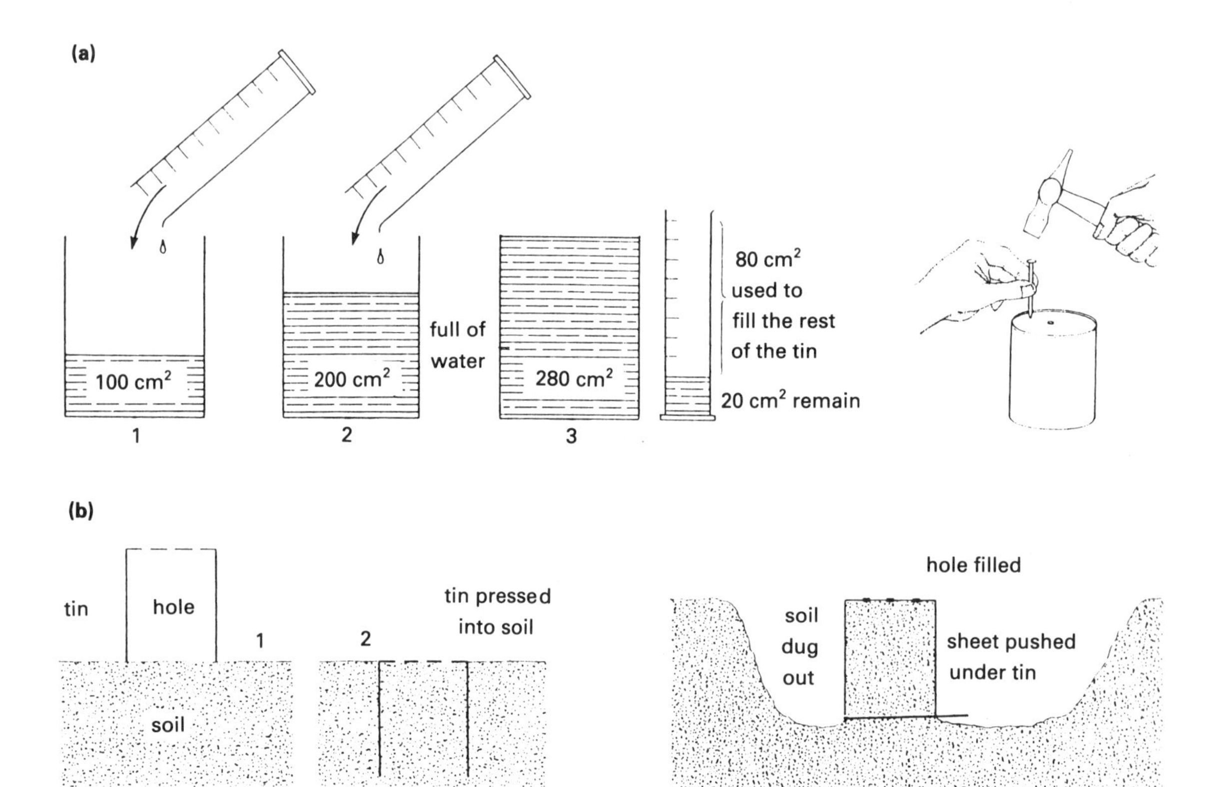

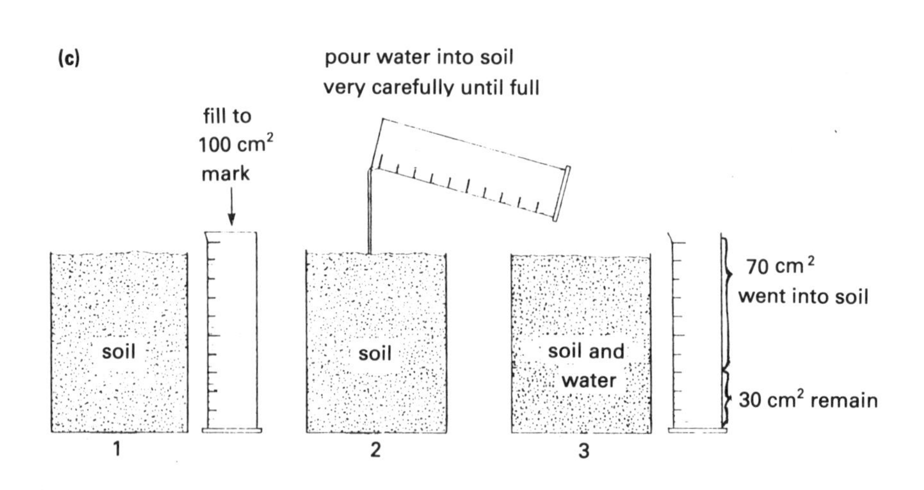

Figure 2.13 Experiment to find out how much air there is in soil

push the tin, open end down, into the soil so that it is completely filled. Do not select a path or other hard ground for this.

7 Seal up the holes in the bottom of the tin using sticky tape, or clay. Scoop away the soil from around the tin, but not underneath, then slip a piece of hard cardboard or thin wood under the tin to hold in the soil. Turn the tin over and stand it upright.
8 Fill the measuring cylinder once again to the top mark with water (d). Record this in Table 2.3.
9 Slowly pour the water on to the soil in the tin until it will not hold any more. Record in Table 2.3 the level of water remaining in the cylinder (e).
10 Subtract this figure from the figure at the top mark of the cylinder. The result (f) will give you the volume of air that the soil held and which has now been replaced by water.
11 The air contained in your soil can now be stated as a percentage by completing the calculation above.
12 Write down in your exercise book how you carried out this test. Discuss the results with your teacher (Figure 2.13).

Organic matter and soil

We have seen that soils contain both organic matter (from decaying plants and animals) and inorganic matter (rock particles and water). Organic matter is very important in soils, and land that grows good crops will usually contain a lot of organic matter. Good farmers will try to add organic matter to their soils.

When plants and animals die in the soil they are broken down by lots of different creatures; a lot of this work is done by bacteria, which are far too small for you to see. In the tropics it does not take much time for dead animals and plants to break down but, in colder countries, this can take several months.

Organic matter is important to soils for several reasons:

- It improves the fertility of the soil by providing food or nutrients to growing plants. As organic matter largely consists of decaying plant and animal remains you can see that it will hold many of the nutrients required by growing plants As organic matter rots down, these vital nutrients are given up to the growing crop.
- It is able to improve soil structure; this is particularly true in clay soils. The particles of clay are tiny and tend to stick together. Organic matter helps to keep them apart so that water can pass through; such soils become well-drained and fertile.
- It helps the soil to hold more moisture; this is very important in sandy soils. Sandy soils do not hold enough water, but organic matter is able to do so. In this way more water is available to plants and the soil is able to produce more crops. Organic matter binds the loose soil particles together.

Organic matter added to the soil is known as an organic fertiliser or an organic manure. Land which has plants grown on it every year will slowly lose its ability to support crops, due to the loss of organic matter. Farmers know this and try to ensure that it is replaced regularly. This is a lesson you must remember so that if you become a farmer your land will stay in good condition and support your crops and your family. In Chapter 3 we will look at the different types of organic fertiliser a farmer may use.

ACTIVITY 7

How much organic matter is there in soil?

This activity is best carried out as a class exercise or in groups. Your teacher will explain to you how to use the scales and what safety precautions you must take. Your teacher will provide:

- a sample of very dry soil (dried gently in an oven would be ideal)
- a small metal or fireproof dish
- a set of accurate scales
- a burner
- tongs, tripod and gauze

Copy down the following table into your exercise book.

Table 2.4 How much organic matter is there in soil? (Activity 7)

Weight of soil + dish	(b)	___ g
Weight of dish	(a)	___ g
Weight of soil	(b – a) = (c)	___ g
Weight of soil + dish before heating	(b)	___ g
Weight of soil + dish after heating	(d)	___ g
Weight of organic matter	(b – d) = (e)	___ g

Percentage of organic matter in soil sample = weight of organic matter (e) ___ g divided by weight of soil (c) ___ g multiplied by 100 = ___%

1 Weigh a small metal or fireproof dish. Record the weight in your notebook (a).
2 Place a sample of the soil in the dish and weigh the dish again. Record that weight too (b). Subtract weight (a) from weight (b) to give you the weight of the soil sample (c).
3 Heat the dish over a burner or other heating device for about one hour (Figure 2.14). You will notice that the organic matter has a distinct smell as it burns away. When the dish is cool weigh it again. Record that weight in your exercise book (d).
4 Now subtract the weight (d) from weight (b); this will give you the weight of organic matter that was burnt off by heating.

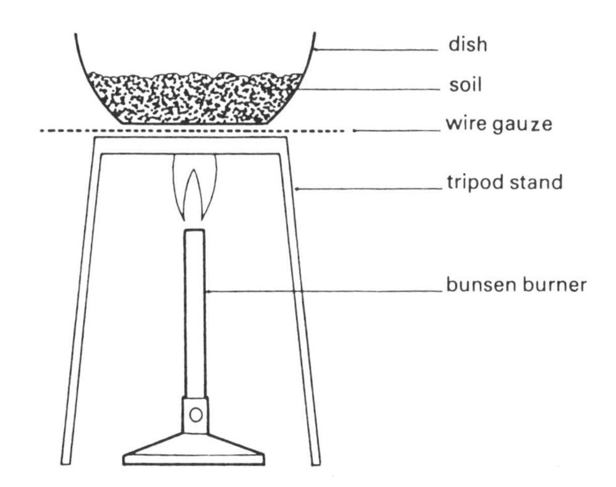

Figure 2.14
Experiment to find out how much organic matter there is in soil

5 You can now express the amount of organic matter in the soil as a percentage by completing Table 2.4 in your exercise book ((e) divided by (c)×100=____%).
6 Write up how you did this test in your exercise book. What was your result? How did this compare with others in your class? What other soils might give higher or lower results?

The structure of the soil

You will remember that we used the term 'soil structure' to describe the arrangement of particles. This is not the same as 'soil texture' which you remember we used to describe the type of particles that make up a soil. Let us look more closely at soil structure.

We have already discovered that soils with a lot of clay particles or organic matter are slightly sticky and that sandy soils are not. When clay and organic matter mix together with sand they form clumps (crumbs). A soil with a good 'crumb structure' allows water to pass through and allows air to fill the spaces

Figure 2.15
Organic matter spread on soil helps to maintain a good crumb structure

between the crumbs; these soils also retain enough water, which is then available to plants.

A soil with a good crumb structure is a fertile soil and one which a farmer will try to maintain on his farm. By working his soil carefully, adding organic matter and growing a variety of crops, a farmer can keep a good crumb structure in his soil (Figure 2.15).

Soil profile

A soil profile is a vertical section (upright view) through an undisturbed soil. If you visit the site of a new road, or any place where deep holes are dug in the soil, you will be able to see layers of different colours. If a new well is being dug in your village go and look at the sides of it when it is about two metres deep. Each soil type has its own special profile; they are not all the same. Agriculturalists are interested in studying soil profiles to find out about the agricultural possibilities of the soils.

A profile is divided into horizons (layers) starting at the soil surface and moving down (Figure 2.16). The layers are given the letters A–D.

A topsoil
B subsoil
C partly weathered parent material
D underlying rock

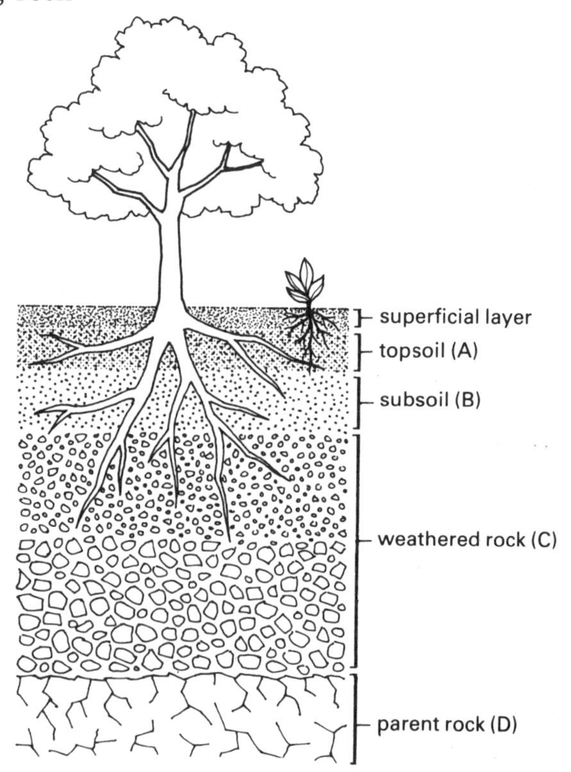

Figure 2.16
A soil profile

The A horizon is used by farmers for growing crops and is known as the 'topsoil'. The topsoil contains most of the organic matter, minerals and water used by crops and it is in this area that you will find most of the plant roots. Farmers try not to disturb this top layer of soil by mixing it with the lower layer. While tractors with heavy ploughs are able to plough to a greater depth, this is not usual practice.

The B horizon is unlikely to contain much organic matter but it will hold a lot of minerals. Plant roots do get down to this level looking for nutrients, water and support.

The C horizon contains partly weathered rock and may have some subsoil in the upper layers. These rock pieces may have been transported from a long distance away (before the A and B layers were formed) or they may have come from the parent rock below. Only the roots from woody trees and shrubs are likely to get down to this layer.

The D horizon is the parent rock (rock from which the soil was formed) or underlying rock. It may be many hundreds of metres thick.

The importance of water in agriculture

In most of the practical activities you have done you will have noticed that water played some part. From this you will have guessed that water is essential for agriculture! Water is vital for both plants and animals.

In Sierra Leone we can always expect to have rain during the year and this rainfall is the source of all the water we use. Think of all the uses we have for water.

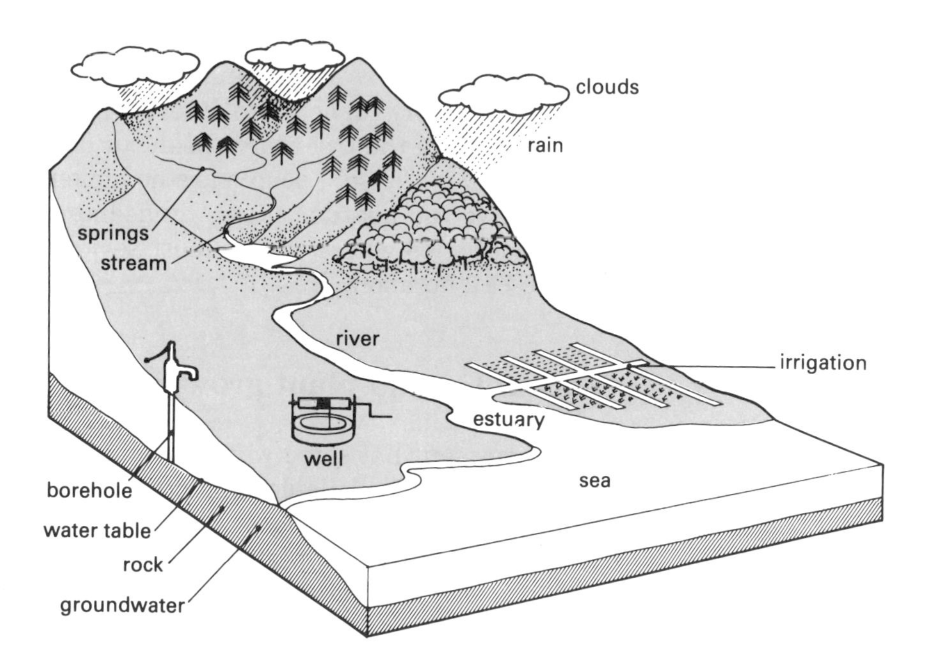

Figure 2.17
Sources of water

The water we raise from village wells has soaked through the soil and lies above a rock or clay layer which does not allow it to pass through.

Much of the rainwater that seeps through the soil finds its way into underground streams which may come to the surface as springs. Alternatively, rainwater will run off the surface of the land. The water from springs and surface run-off joins up to form streams and rivers which flow to the sea. Rivers may be used to irrigate the swamps, and low-lying land used for growing rice (Figure 2.17).

Figure 2.18
A balance between forest and farming ensures rains with no erosion (left)
Figure 2.19
A deserted village in Mali where the land has turned into desert (right)

We must not take water for granted; it is much too valuable. We must try to use it wisely and not pollute (dirty) our water supplies. It has been shown that our rainfall is closely linked to the presence of our forests and bush land. We must make careful use of our land and keep the natural vegetation cover, to make sure that the rains will come each year and that the rains do not wash away the soil (Figure 2.18). In many parts of Africa the people have cut down trees and the rains have failed to come; the land has quickly turned into desert (Figure 2.19). In many other parts of Africa the trees have been cut down and the rains have washed away the soil; people are no longer able to farm on such land. Have you heard of these things happening? Can you name some countries where this is most serious? What do you think people can do about it?

Soil water and plant growth

We have seen that some water is held around soil particles and that this water is available to plants to take up through their roots. Once inside the plant, water has several very important roles to play in the life of the plant.

First we must understand that water is able to dissolve (melt away) many substances; this includes the surface of the soil particles. When this happens the minerals from the soil particles move into the soil water (soil solution) and may be used by the plant as food.

The plant's roots come into contact with this soil solution and take in some water and selected minerals. These plant nutrients can then move around inside the plant, go to where they are needed, and are made into new plant material.

Water is also necessary to keep soft green plants standing upright. Without water, plants tend to fall over (wilt). Have you ever seen plants wilting? What happens if you give a wilted plant some water? What happens if a plant is allowed to wilt too much? Perhaps you can now see why farmers have to water their crops so much!

We have all noticed that when we stand in the hot tropical sun we get hot and start to perspire (sweat); our skin gives out water which keeps us cool. Plants are no different! Water is also essential for cooling plants. Plants give off moisture through their leaves (a process called transpiration); this keeps them cool and also helps draw up nutrients from the roots.

Without a proper supply of water, crops will not be able to grow properly. If there is not enough rainfall, farmers must supply the crops with water from other sources; this is called irrigation.

Water for irrigation may be collected from many sources, including water collected from roofs, wells, boreholes, streams and rivers, dams and reservoirs. The water may be carried to the crops by hand or moved along ditches or pipes. Irrigation is costly and requires a lot of effort, so we must not waste the water we use.

ACTIVITY 8 Do plants need water?

Comparing plants kept in a dry soil with plants kept in a wet soil is a very simple test which can be carried out as a group activity.

Dig up two square pieces of damp soil (10 cm × 10 cm) which have grass growing on them. Put each piece in a polythene bag with the sides rolled down. Put both bags side by side in a sunny place; a windowsill or a table would be ideal. Do not water them.

Label them A and B. Examine them once in the morning and once in the afternoon. How long is it before the grass goes limp and bends over (wilts)?

When they have wilted, add some water to bag A only. How long is it before the grass in this bag looks strong again? Leave bag B for four days and then water it. Does bag B recover or has the grass died? If bag B recovers, does it look exactly like bag A?

What conclusions can you come to about the need of plants for a regular supply of water?

Some key words and terms in this chapter

Transported soil A soil which was not formed where it is now found; it may have been moved by wind or water to its present place.
Carbon dioxide A gas which is found in the air. It may be absorbed by rainwater to make a weak acid.
Acid A corrosive liquid.
Biological weathering The breaking down of rocks by the action of plants and animals, to form soil.
Mechanical or physical weathering The breaking down of rocks by wind and water, to form soil.
Chemical weathering The breaking down of rocks by weak acids contained in rainfall, to form soil.
Soil texture The 'feel' of a soil, which depends on the different particles present.
Soil structure The arrangement of soil particles within a soil.
Organic matter Plant and animal matter, whether living or dead, which is contained in a soil.
Loam Soil which is made up of roughly equal parts of sand, silt and clay.
Pore spaces Spaces between soil particles, which are filled with air and/or water.
Soil profile A vertical cutting through a soil to show its different layers.
Horizon A layer or band of soil, as seen in a soil profile.
Parent rock A rock which, through weathering, gives rise to a soil.
Pollute To contaminate or dirty something; as in pollution of the environment or pollution of water.
Transpiration The process in which plants give off water from their leaves to stay cool.

Exercises

Multiple choice questions

1 A loamy soil consists of particles of:
  a clay and loam
  b sand and clay
  c sand, silt and clay
  d clay and organic matter

2 The term biological weathering applies to which of the following?
  a the action of animals and plants on particles in the soil
  b the production of soil from volcanic eruptions
  c the action of acids from the air on rocks
  d the action of wind and running water on rocks

3 Which of the following soils is the best for crop production?
  a sandy soil
  b clayey soil
  c loamy soil
  d clay-sand soil

4 The type of soil found in an area is mostly influenced by:
  a the type of vegetation
  b the parent material
  c shape of the land
  d the crops grown

5 The relative amount of sand, silt, clay and organic matter in a soil is called the:
  a soil consistency c soil profile
  b soil structure d soil texture

6 A light brown soil was found to be very sticky when wet, and formed a ribbon when rolled between the fingers and the thumb. To which class does the soil belong?
  a loam c sand
  b clay d sandy loam

Puzzles

7

C	M	L	A
A	L	W	O
D	N	A	S
E	R	T	Y

Copy this diagram into your notebook.

Can you find the two words 'clay' and 'sand' in the diagram. The letters are in straight lines but may be written diagonally backwards. When you have found them draw a ring in pencil round the two words.

Now you have nine letters left. Arrange them into two words, one of which is essential in the soil for plants to be able to grow, and the other is a name for a mixture of sand, clay and silt.

8 Copy the diagram into your exercise book and complete the following clues.

Across

1 Describes soil material made up of dead animals and plants (7)

2 How a silty soil feels when rubbed between your fingers (8)

3 A layer seen when a pit is dug to give a vertical view of a soil (7)

4 A soil found in an estuary by the sea (4)

Down

5 A soil which has been moved by water or wind (4)

6 The term given to a vertical view of the soil (11)

Missing words

9 Complete the following sentences in your exercise book by selecting the most appropriate words from the list below.

The forest trees and bush help clouds to form and this brings ...... to our countryside. When there is not enough ...... in the soil the crops may ...... Farmers may then have to ...... their crops. The water added to the soil fills the ....; some ...... away and some is held around the soil ...... and is available to plant ......

drains	**particles**	**pore spaces**
irrigate	**roots**	**wilt**
rainfall	**moisture**	

Points for discussion

10 What different soils are found in your area? What type of soils are they? Where exactly are these soils? Which soils are good for farming? Who has farms on the best soil?

11 Ask the old people in your village what the countryside was like when they were young. Were there more trees? Do they think the climate has changed over the years? How has it changed? What might have caused these changes?

12 Do you see any evidence of soil erosion in your area? What has caused this? How could it be stopped?

13 Where do the people in your district get their water from? Is the water pure? Do farmers irrigate any of their crops? When and how is this done?

CHAPTER 3 Plants and soil fertility

Introduction

We studied the soil in the previous chapter. It is now time for us to find out how plants are able to use it in order to grow. In this chapter we will look very simply at how plants work and how farmers can keep their soil in good condition to allow their crops to grow to their full capacity. We will discover how farmers can make the soil produce more by adding fertilisers and manures.

The growing plant

We have seen in the previous chapter that the soil is capable of holding the nutrients (food) and the water that plants require, and that plants are able to take in selected nutrients through their roots. Now we must examine the plant in detail to understand how it does this and how it is able to grow. As agriculturalists, it is most important that you should study living plants, but before you do so, we will draw one and explain how it works.

Look at the diagram of the plant on this page. Study the names of the parts and try to remember them. It will help you when visiting farms.

Parts of the plant

ACTIVITY 1

Your teacher will provide you with a plant that has been carefully dug up; a maize plant like the one in Figure 3.1 would be ideal. Draw it in your exercise book and label the parts. Does it look like the plant in Figure 3.1?

The roots

First look very carefully at the roots of the plant. What do you see? The roots form the underground part of a plant and seem to be like the branches of a tree in reverse! Follow one root down from the base of the stem to the very end. Did you notice that it got thinner and thinner? How long was it? How many times did it branch?

Here are some of the jobs that roots do:

- They hold the plant in the soil. Without roots plants would just fall over! Can you imagine how strong the roots of a huge forest tree must be to hold the tree up in a strong wind! If you have a maize plant look carefully at the roots that grow out of the stem just above soil level; these are special supporting roots called 'prop roots'. Not many other plants have them. Why are they called prop roots?
- They take in water and nutrients; they also pass out waste materials into the soil. In the last chapter we found out that all these exchanges take place in the soil solution (the water surrounding the soil particles). Water and nutrients are taken in through the smallest of roots, called root hairs. Can you see any tiny white root hairs?
- Some plants use their roots to store food. You will have seen that cassava, for example, has swollen roots. These roots hold stores of food in the form of starch which allows the plant to begin to grow in the following year. Yam and sweet potato are other examples of crops that store food in their roots (Figure 3.2).

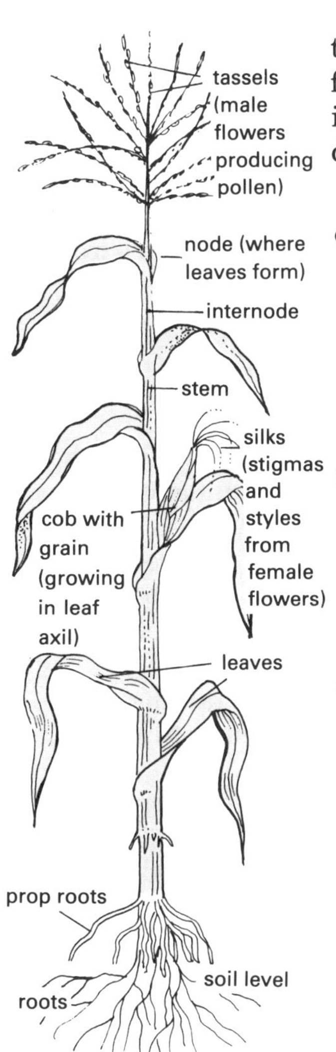

Figure 3.1
Maize plant

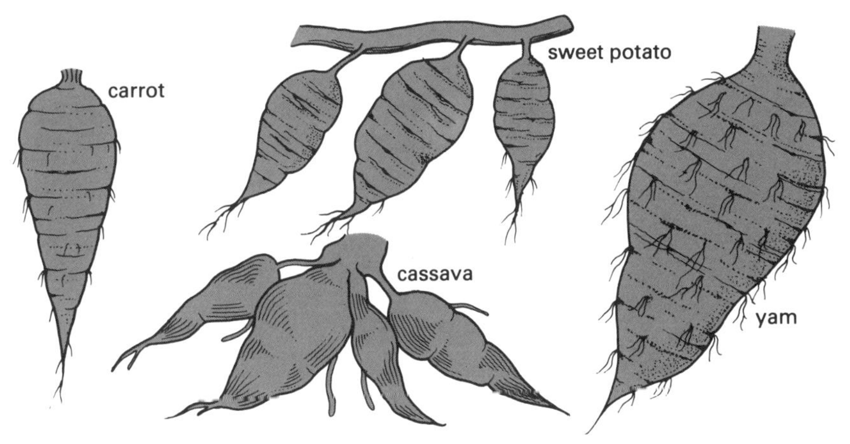

Figure 3.2
Some of the plants that use their roots to store food

Can you name any other plants that provide us with useful roots? Be careful! Not all the things we harvest from below the soil are necessarily roots. We shall see that things like Irish potatoes are, in fact, underground stems! Do you know of any forest trees that supply us with medicine or dye from their roots?

Let us look at the colour of the roots of your plant. What colour are they? They are either pale brown or white in colour but you will notice they are **not** green. If you have a maize plant you will see that the prop roots are green when they grow from the base of the stem until they reach the soil. Why are roots not green?

This is a question we will answer more fully later in this chapter but, for the moment, we can say that it is because they are not in the light.

We have said that plant roots are like the branches of a tree in reverse. This is because roots contain a network of pipes which link up to carry nutrients and water from the soil to the stem and leaves. There is also a network of pipes that bring food down from the leaves to the roots.

Stems, stalks trunks and branches

These are all words that describe parts of a plant, other than the leaves, that are above the ground. What do you understand by these words?

The term trunk is used when we are talking about a tree which has a large, hard, woody support. We would not talk about the 'trunk' of a maize plant would we? Why not? Look carefully at your maize plant. What do you see?

Stems and stalks are words we would use when talking about smaller, softer, greener plants. We would, for example, talk about the stem or stalk of a maize plant.

Branches are formed when a trunk, stem or stalk splits up into smaller divisions.

All trunks, stems, stalks and branches are supported by the roots of the plant and provide a framework for the plant. This framework has several jobs. Can you think what they might be?

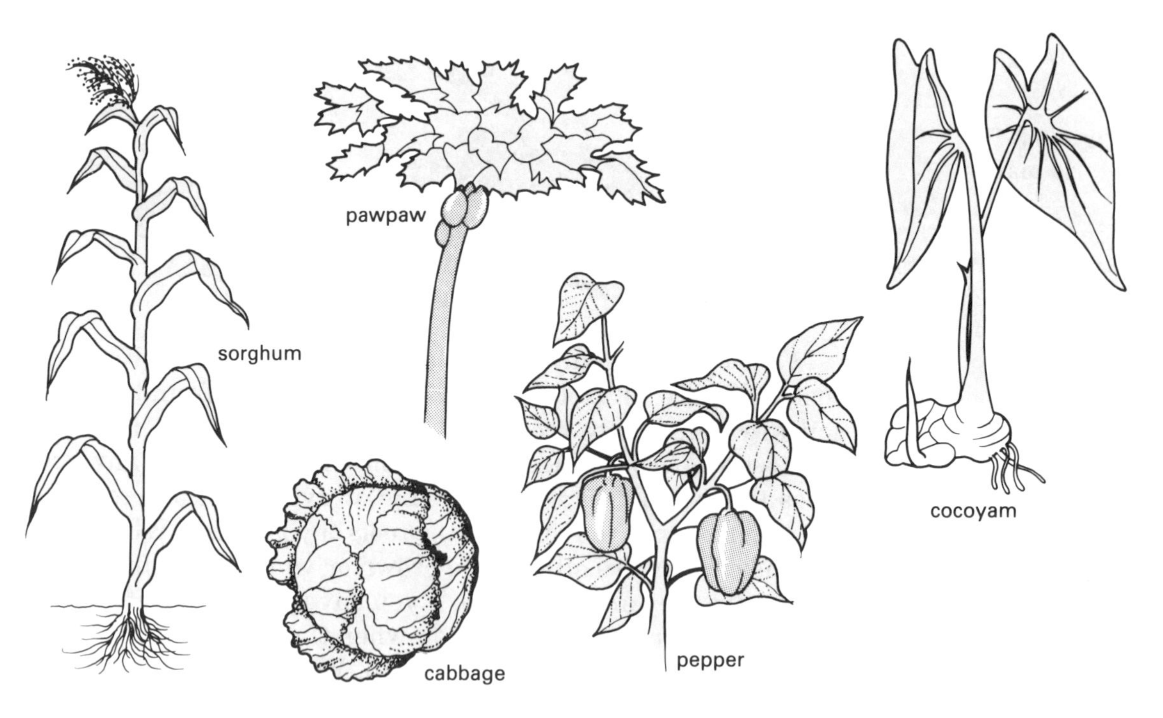

Figure 3.3 Every plant has its own special arrangement of leaves, branches and stems

Figure 3.4
Sugar cane stores sugar in its stem

This is what trunks, stems and stalks do:

- They hold the leaves, flowers and fruits up into the air. Each plant has its own special arrangement of parts to do this (Figure 3.3). Look carefully at different plants around you and see how they are arranged. It is very important that plants' leaves are held up to the sunlight and that their flowers or fruit are held up in the best possible way. Look at a forest and see how all the trees are trying to reach the best possible place in the sunlight.
- They transport water and nutrients up to the leaves and move food in the form of sugars and starches down to the rest of the plant.
- Some special stems may become enlarged and store food for use by the plant in the following season. You will probably know that sugar cane stores sugar in the stem (Figure 3.4). It may surprise you that some of these special storage stems may be formed **under** the ground. Irish potatoes are, in fact, special underground stems.

As you walk around your home, look at different plants around you and notice how they all differ. Some are tall and woody, some are small, soft and green.

Can you think of any plants that provide us with useful stems? What about mahogany, rubber, sugar cane? How are they used?

Leaves

The special purpose of leaves is to manufacture complicated foods from simple substances. The plant needs these complicated

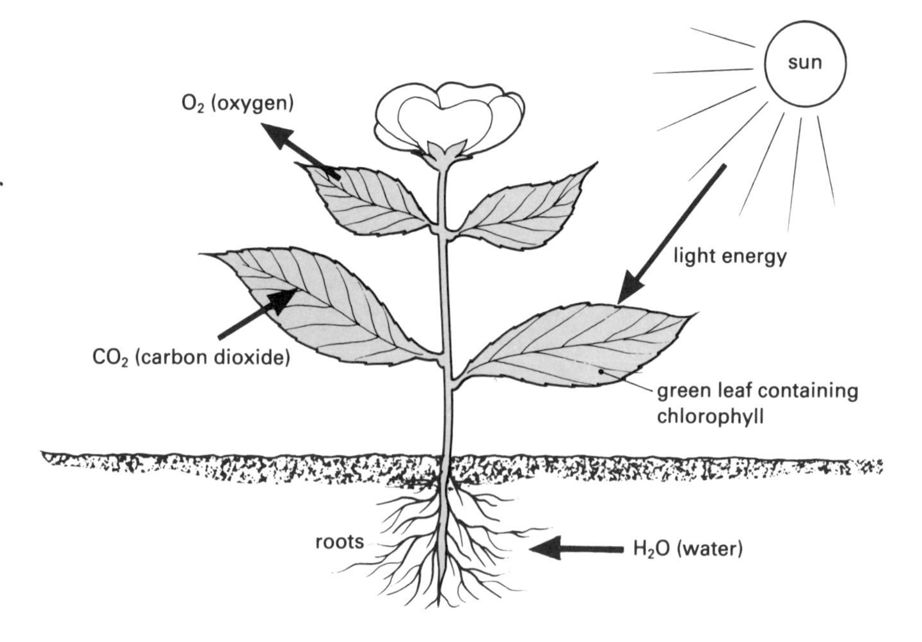

Figure 3.5
The process of photosynthesis uses water from the roots, carbon dioxide from the air and energy from sunlight to produce sugars

substances if it is to grow. You can think of the leaves as the plant's factory! Let us find out how this factory works!

First we must understand that leaves are able to breathe and, in particular, are able to take carbon dioxide from the air. Secondly we must know that, effectively, all leaves are green. This is due to the presence of a substance called chlorophyll which, as we will see, is the key to the whole factory process! Thirdly, we must remember that the plants draw up a solution of water and mineral salts from their roots.

The manufacturing process can only take place in daylight! In sunlight, the chlorophyll in the leaves is able to trap energy from the sun. By using carbon dioxide from the air and water from the roots the leaves are able to use the trapped energy to make sugar. This process is known as photosynthesis (Figure 3.5). Roots do not contain chlorophyll, which is why they are not green but white.

However plants, like humans, need a balanced diet if they are to grow and thrive! Sugar alone is not enough! Plants must make other substances. The leaves use the sugar made during photosynthesis and combine it with minerals from the roots to produce a whole range of substances required for growth; these will include proteins, fats and oils. Just think of the things we get from plants – timber, rubber, palm oil, cocoa – all of these made from air and a few simple minerals from the soil! You can see that leaves really are very clever factories!

If you were to look at the underside of leaves very carefully with a magnifying glass you would see that they were covered with tiny holes. These holes are called stomata (Figure 3.6). These holes allow the plant to 'breathe' and to give off water. We have seen that plants need to use air in photosynthesis. Plants must also give off moisture to stay cool and to draw up minerals from their roots. This is called transpiration.

Once again you are asked to look closely at the plants around you! Do you notice how plants arrange their leaves? They are all trying to catch as much sunlight as possible. Look at all the different shapes of leaves. Draw some crop plants in your exercise book to show the leaf arrangement, like the examples shown in Figure 3.3.

There are lots of plants that provide us with useful leaves. See how many you can name!

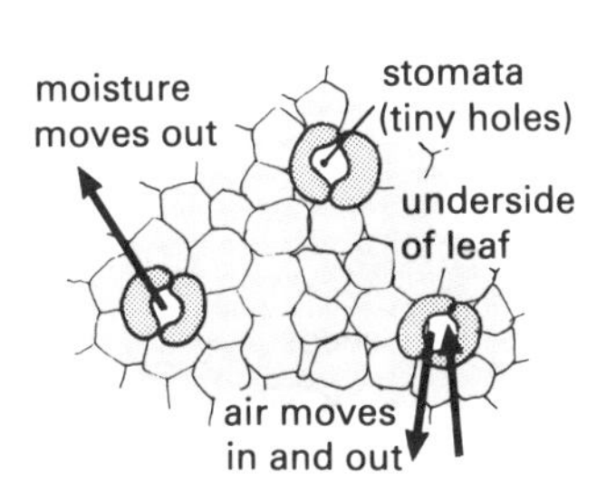

Figure 3.6
Air and moisture pass through tiny holes in the lower surface of leaves called stomata. This is what stomata look like through a magnifying glass

Flowers, fruits and seeds

Plants, like animals, must be able to reproduce so that they can go on from generation to generation. Plants have developed a number of ways of doing this. Some plants produce flowers, fruits and seeds as their way of reproducing themselves.

The flower is the first stage in this process. Once a flower matures and is fertilised, a fruit is formed which will contain one or more seeds. After the seeds ripen they will be able to germinate (sprout) and produce new tiny plants (Figure 3.7).

Fruits and seeds are very important to farmers. A great number of the crops we grow are harvested for their fruits or seeds. Figure 3.8 shows a few examples. Can you make a list of 20 more? It should not be hard to think of them!

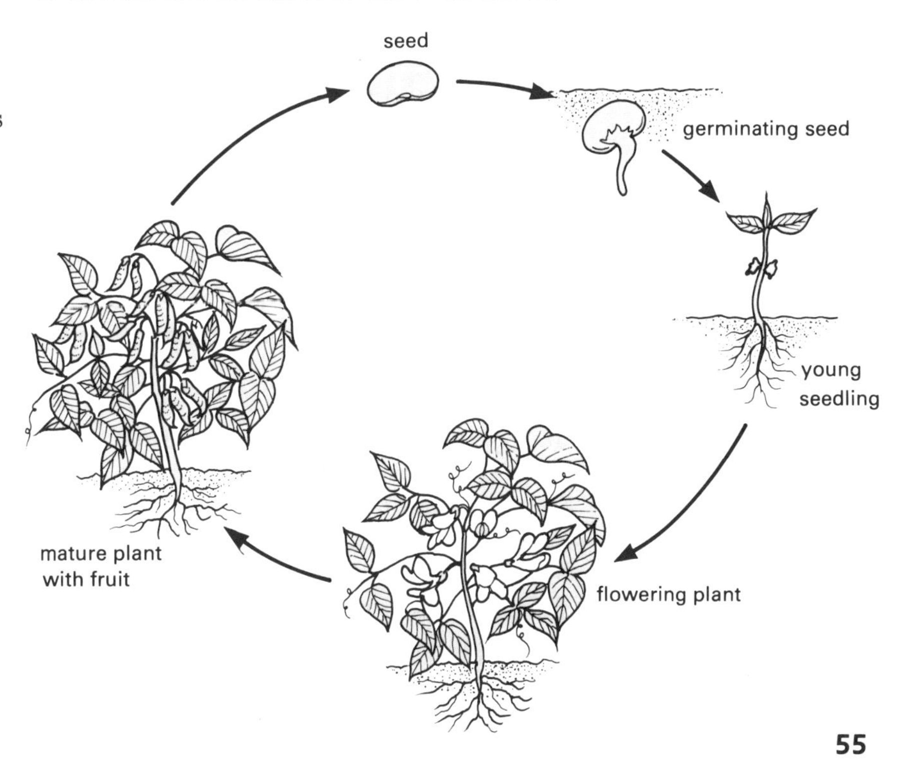

Figure 3.7
Life cycle of a flowering plant. The flower produces the fruit, which produces the seed, which in turn grows into a new plant

We must remind ourselves that, once again, it is the plant, through its roots, stem and leaves, which supplies the flowers, fruits and seeds with all their needs.

Note: Be careful not to confuse the word fertilisation in plants (the final stage of the pollination of a flower) with fertilisers (plant nutrients) which are applied to the soil.

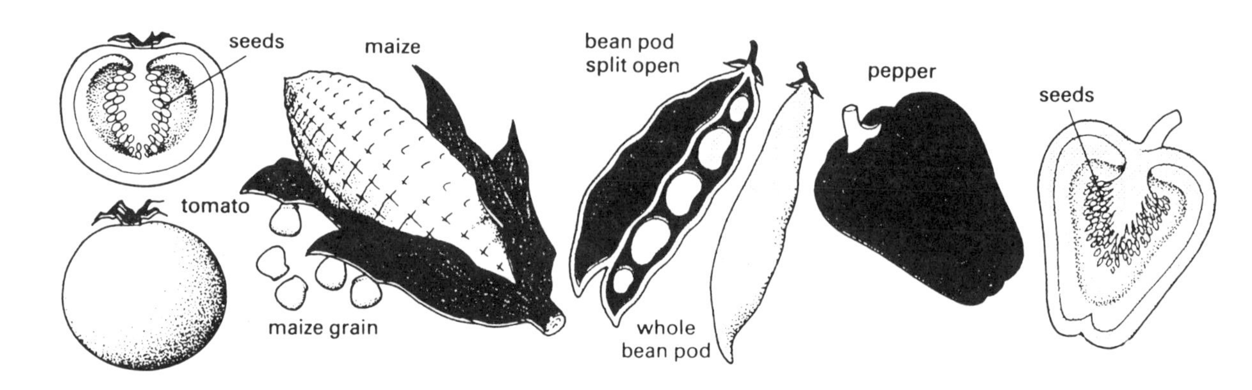

Figure 3.8
Some of the fruits and seeds grown in Sierra Leone

Water in the plant

Have you noticed that water was always being mentioned in our discussions on roots, stems and leaves? You will have gathered that water is as essential for plants as it is for humans and other animals!

In this section we will look at the important jobs that water has to do in plants; some of these will already be known to you:

- Water is needed to transport substances around the plant. Water containing dissolved mineral salts is taken up by the roots and moved up through the plant to the leaves. Substances made in the leaves are then transported to all parts of the plant. Materials are moved around the plant through fine tubes called vessels.
- Water fills the cells making up the fleshy part of plants, keeping them tough and rigid. This keeps the plant standing upright. If plants have to go without water, they quickly wilt (lean over) and die.
- Water evaporates through the stomata on the leaves helping to cool the plant. Water evaporating from the leaves helps to 'pull' up more water and nutrients from the roots (Figure 3.9). Just think how high a forest tree has to raise the water from its roots to its leaves! A huge tree can loose many hundreds of litres of water through its leaves every day.
- Water takes part in all the life-giving processes in a plant. We have seen, for example, that in photosynthesis, water combines with carbon dioxide to produce sugar.

Figure 3.9
Diagram to show the path of water moving through a plant from the roots to the leaves

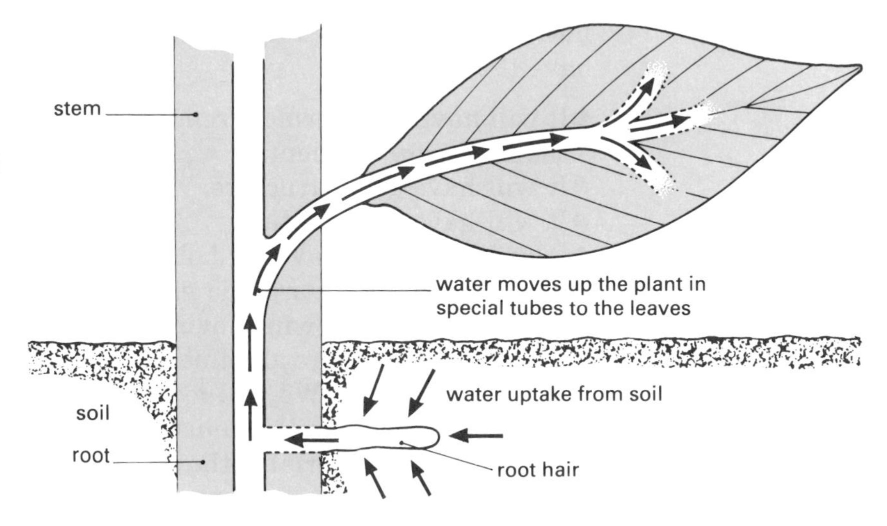

We have just been studying how a plant works and how farmers use different parts of the plant. In the next part of this chapter we will find out how farmers make sure that the soil is able to supply their growing crops with everything they need.

Soil fertility

If crops are to grow as well as possible, they must be able to get a steady supply of all the nutrients they need. These nutrients must be available in the correct amount at the correct time. If a soil can supply all the crop's needs, the soil will be fertile and a farmer's crops should grow well (Figure 3.10).

We can define soil fertility as the quantity of available nutrients present in the soil and the ease with which these

Figure 3.10
A fertile soil provides a good growing environment for crops, such as rice

nutrients can be taken up by plants. A fertile soil will therefore have the following:

- It will have been formed from a parent rock able to supply all the necessary nutrients.
- It will have good structure.
- It will have good texture.
- It will have a high water-holding capacity.
- It will have a high organic matter content.
- It will be a good growing environment for the crops. Some soils are acid and some are alkaline; most crops seem to prefer a very slightly acid soil. (We will look at this in later books.)
- It will have lots of soil organisms. These tiny creatures are able to break down material in the soil and supply the growing crops with nutrients.
- It will be a rich dark colour. A dark colour usually indicates a fertile soil. However, this is not always the case! The soils around the Njala and Kone districts are light brown but are very fertile.

A soil that does not have the essential plant nutrients or is unable to give them up to plants is said to be infertile.

Look at the soils in your area. From their appearance, do you think they are fertile or infertile? Why do you think this? Do the crops in your area grow well? Does this confirm your ideas about the fertility of soils from your district?

Importance of soil fertility

If farmers are to be successful we can see that they must keep their soils fertile. Farmers understand this very well and they take steps to make sure that their soil is kept in good condition.

We know that plants draw up minerals from the soil. When crops are harvested and carried off the fields these minerals go with them. If the farmer does not replace the minerals he has taken away, the soil will become less and less fertile; this will happen slowly over a number of years. If a soil is to remain fertile, the minerals that have been removed must be returned. Ideally a farmer would wish to add more nutrients to his soil than he has taken away so that, over the years, soil fertility will gradually build up.

Traditional farming and soil fertility

In our traditional agriculture, a farmer clears a patch of forest or bush in order to grow crops. The farmer usually burns the trees that grow there (Figure 3.11); this releases the nutrients from the trees and bushes and makes them available for the crops which are about to be grown. The farmer will probably use this area for only five or six years. During this time the yield of the crops will fall as the soil becomes poorer and poorer. Eventually the farmer must abandon the area and move off to somewhere else. The bush gradually returns and the soil has a chance to recover. The farmer may return to the same patch of land 20 or 30 years later. Resting soil in this way is known as 'fallowing' and is an important way of restoring soil fertility. The farmers who practise this traditional 'slash and burn' farming understand all about soil fertility!

Figure 3.11
'Slash and burn' agriculture is a traditional method of farming in Sierra Leone

As the population in Sierra Leone increases and less land is available, we cannot rely on this traditional form of farming. Farmers must grow crops on the same patch of ground year after year and so they must look after their soil fertility.

Maintaining soil fertility

We should already have quite a few ideas as to how farmers can maintain or improve their soil fertility. We have explored some of the ideas in Chapter 2 and we have met another in this chapter! Can you remember what they were?

Here is a list of the ways in which a farmer can maintain soil fertility:

- *Adding organic matter to the soil.* You will remember that organic matter can include farmyard manures, composts and green manures.

- *Fallowing.* By resting the soil a farmer allows it to recover its fertility. Fallowing allows plant material to decay in the soil, thereby increasing soil fertility.

These are the ways of improving soil fertility that were already mentioned. Did you remember them? Here are some more:

- *Applying inorganic or artificial fertilisers to the soil.* Farmers can buy manufactured chemicals to mix with their soil. These chemicals provide crops with an exact amount of nutrients. We will look at these fertilisers later in this chapter. Have you seen a farmer putting an inorganic fertiliser on his soil?
- *Soil erosion control.* In many parts of Africa, farmers are not looking after their soil. They are growing crops on very steep land, allowing rain and wind to carry away the soil. If people are to continue farming their land for generations they must make sure that the soil is not lost. There are many things farmers can do to stop erosion; we will consider these in later books. Planting trees and growing crops on terraces (flat beds) built on sloping land are two examples of soil erosion control (Figure 3.12). Have you ever seen soil erosion? Where was it? How was it caused? Can you think of anything farmers do to stop their soil from being lost?

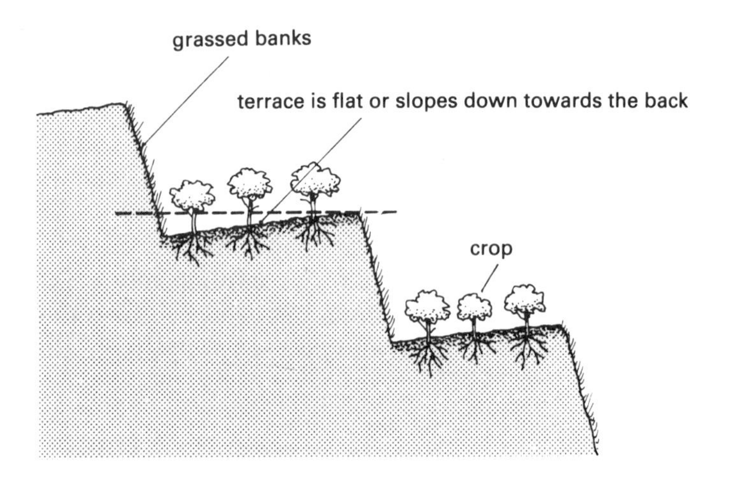

Figure 3.12
Cross-section of a terrace. Terraces are built on sloping land to prevent soil erosion

Fertilisers and plant growth

A fertiliser is another word for a manure and is any substance added to soil to supply the nutrients needed by plants for their healthy growth.

If a plant is to grow really well it must have a 'balanced diet' of plant nutrients. Some nutrients are required in greater amounts than others. If one or more of these nutrients is lacking the plant will not thrive. It has been found that plants need three nutrients in greater amounts than any others; these are substances called nitrogen, phosphorus and potassium (sometimes called potash) and are given the letters N, P and K for short! Farmers often think about fertilisers in terms of the amount of N, P and K contained in them. We will see that this is very important when we talk about fertilisers later in this chapter.

Types of fertilisers

From this and the previous chapter we have already found out quite a bit about fertilisers. Let us now look at fertilisers in more detail.

Fertilisers may be divided into two groups:

- organic fertilisers
- inorganic fertilisers

Organic fertilisers

Figure 3.13
Construction of a compost heap

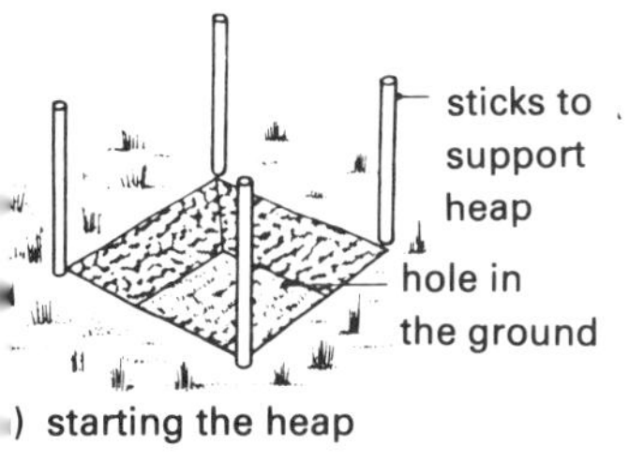

) starting the heap

layers of grass, crop residues and animal manure
stick to test temperature

) the heap after being built

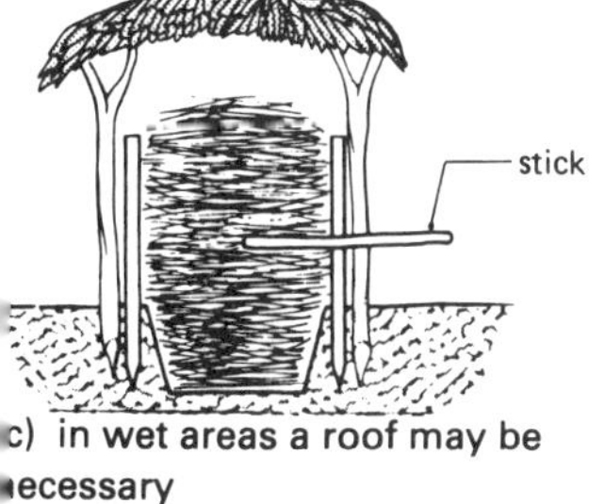

c) in wet areas a roof may be ecessary

Organic fertilisers are those that are formed from plant or animal remains; when added to the soil they increase the organic matter content of the soil and give the farmer many benefits, including improvement in soil structure.

Organic fertilisers (or organic manures) are very important to agriculture in Sierra Leone. You will see that they are usually prepared on the farm and are cheap and effective.

Let us look at the sources of organic fertilisers:

Compost
When leaves, plants and unwanted vegetation are piled into heaps or placed in pits, they rot down. This is helped by the heat that builds up inside the heap and is caused by a process which is the same as happens when organic matter is broken down in the soil. The material that eventually forms (compost) is dark brown, moist and smells of clean earth. Good compost is made up entirely of organic matter and is dug into the soil to help plants grow.

Figure 3.14
A farmer spreading farmyard manure on his land (right)

Farmyard manure
Cattle, sheep, goats, poultry and other animals produce manure which is one of the best organic fertilisers. Farmyard manure is gathered up by farmers, spread on their land, and dug into the soil (Figure 3.14).

Green manures
Some farmers grow crops specially as organic manures; these are called green manures. Crops like Crotalaria grass are ploughed into the soil while they are still growing to provide nutrients for the next crop. In Sierra Leone this is only done on large farms that have a lot of machinery.

Fire ashes
We have seen that in our traditional agriculture, the bush is cleared and trees and vegetation are piled into heaps and burnt. This has the advantage of giving back some nutrients to the soil, particularly potassium, and making the farm clearance quicker (Figure 3.15). It would probably be better if a lot of this material was made into compost!

Figure 3.15
In 'slash and burn' agriculture, the ashes provide nutrients for the next crop. In the picture, healthy crops of rice and maize are growing on newly cleared land

Organic manures and soil fertility
Organic manures provide organic matter for the soil. You will remember that organic matter is broken down by very small creatures in the topsoil. During the process, a dark-brown moist material is formed, which is called humus. It has an open structure and holds many plant nutrients, particularly nitrogen, which are essential for plant growth. Therefore, by adding organic manures to a soil, a farmer improves its:

- fertility
- drainage
- structure
- ease of working

- moisture holding ability
- and it encourages all the tiny soil creatures to become active and more numerous

You can see why organic fertilisers are very important. Have you ever seen a farmer making compost or collecting manure to dig into the soil?

Inorganic fertilisers

Inorganic fertilisers are those fertilisers that do not have their origin in living plants or animals; they are usually made in factories from materials that have been mined from the ground. It is for this reason that they are sometimes called artificial or mineral fertilisers.

While these fertilisers can produce very good yields of crops, they are expensive and not all farmers can afford them. When farmers add inorganic fertilisers to their soils, these usually contain exact amounts of one or more of the main plant nutrients.

Do you remember the three main plant nutrients? They were N, P and K! Do you remember what the letters stand for? Nitrogen, phosphorus and potassium (sometimes called potash)!

Inorganic fertilisers can be classified into two groups:

- straight fertilisers
- mixed or compound fertilisers

Straight fertilisers

Straight fertilisers are those that contain a single plant food or nutrient in a form that plants can easily use. Some examples of straight fertilisers are shown in the table below:

Table 3.1 Straight fertilisers

Name of fertiliser	Plant nutrient supplied	Analysis**
Ammonium sulphate	Nitrogen	21%
Urea	Nitrogen	45%
Super phosphate	Phosphorus	20%
Muriate of potash	Potassium	40 – 60%

** Analysis means the percentage of plant nutrient contained in the fertiliser.

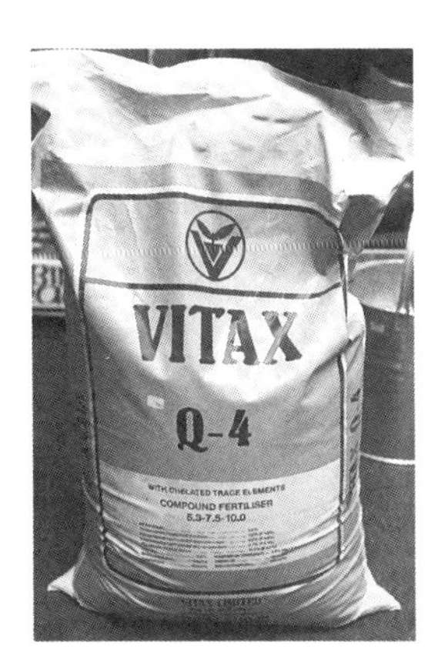

Figure 3.16
A bag of compound fertiliser

Mixed or compound fertilisers

Mixed fertilisers contain more than one plant nutrient.

We have seen that plants are most likely to be short of nitrogen, phosphorus and potassium; these are the nutrients normally given to plants in mixed fertilisers. Mixed fertilisers are made with different amounts of N, P and K to suit different crops and different situations (Figure 3.16).

Applying the right quantity of fertiliser

Farmers will want to know how much fertiliser they are going to apply; like making a cake they must make sure they use the right amount of each ingredient! How do they do this?

If you look carefully on the side of a bag of fertiliser you will see that it says how much of each plant nutrient is in the fertiliser. This is written as a ratio and the numbers show how many kilograms of each nutrient are contained in 100 kg of the fertiliser. For example a bag may say 20N:10P:10K. This means that the bag contains 20% nitrogen, 10% phosphorus and 10% potassium. In fact the bag may say 20N: $10P_20_5$: $10K_20$. Do not let this worry you, it means the same thing! If the bag weighed 50 kg, as indeed most bags do, then it would contain 10 kg of nitrogen, 5 kg of phosphorus and 5 kg of potassium. It's quite easy really!

Fertilisers are normally sold in plastic bags that weigh 50 kg each. It is now possible to buy half tonne bags, but these need special tractor handling equipment to move them.

Other nutrients needed by plants

We have seen that plants are most likely to run short of nitrogen, phosphorus and potassium. Plants, however, need a whole range of nutrients (elements) if they are to grow well.

Nutrients that plants need in quite large amounts are called major plant nutrients and, of course, include N, P and K. Others in this group include carbon (C), hydrogen (H) and oxygen (O), which plants obtain from water (H_2O from the soil) and from carbon dioxide (CO_2 from the air). Also in this group is sulphur (S), calcium (Ca), magnesium (Mg), sodium (Na) and silicon (Si).

Plants need other nutrients in very small amounts; these are called minor plant nutrients or trace elements. Do not think that these are any less important because they are needed in only small amounts; they are vital if the plants are to grow well! This group contains such elements as manganese (Mn), boron (B), copper (Cu), zinc (Zn) and iron (Fe).

A crop which is short of a plant nutrient is said to suffer from a deficiency; this may cause the plant to look sick! It is then said to be suffering from a deficiency disease.

Advantages and disadvantages of inorganic fertilisers

The advantages of using inorganic fertilisers are:

- The farmer can apply an exact amount of plant nutrients.
- The fertiliser is easy to use.
- The farmer can expect to get a good yield from his crops.

The disadvantages of using inorganic fertilisers are:

- They are expensive to buy.
- If the farmer gives too much fertiliser to his plants he can damage them.
- Inorganic fertilisers dissolve in water easily and nutrients can be easily washed out of the soil; this is particularly true of nitrogen. The loss of nutrients in this way is called leaching.
- While these fertilisers add nutrients to the soil, they do not improve the soil structure.

What do inorganic fertilisers look like?

Your teacher may be able to show you some examples of inorganic fertilisers. You will see that they come in a variety of different colours and physical types.

It is important that you should be able to describe them. Here are some words that are commonly used:

- *Powders.* Here the fertiliser is ground up into a fine dust with very small particles. Muriate of potash is a powder.
- *Granules.* These are irregularly shaped particles which may vary slightly in size. Mixed fertilisers are usually granules.
- *Crystals.* Here the particles of fertiliser have a regular structure. Each particle will be shiny and angular. Ammonium sulphate is a crystalline fertiliser.
- *Prills.* These are round shiny white balls of fertiliser, which are the same size and flow very freely over each other. Ammonium nitrate is an example of a prilled fertiliser.
- *Liquids.* Sometimes fertilisers may be applied as a liquid.

Look at different inorganic fertilisers; make a note of their colour, their physical appearance and the amount of plant nutrients they contain. Make notes about them in your exercise book.

Storage of inorganic fertilisers

If farmers are to get the best from their inorganic fertiliser they must make sure it is stored properly; it is much too expensive to allow it to go to waste!

When fertiliser is stored it should be kept out of the sun and rain and raised above the ground on some form of low platform. While fertilisers are usually sold in tough plastic bags which are sealed against the entry of water, the plastic can be torn if the bags are not stored properly and when the bags are wet they are not easy to handle.

If bags are left in a dark building for long periods there is a chance that small animals will damage the lower bags and snakes may hide under them. Stocks of fertiliser should be inspected at regular intervals to make sure they remain in good condition.

Applying inorganic fertilisers

Several methods can be used to apply inorganic fertiliser to the soil:

- *Broadcasting.* This involves the sprinkling of fertiliser evenly over the entire plot of land by hand or machine. A machine for sprinkling fertiliser is called a spreader (Figure 3.17). The fertiliser may then be mixed into the soil. Broadcasting is a simple and cheap way of spreading fertiliser.
- *Top dressing.* This term is used when fertiliser is spread over and around a growing crop. Grass is often top dressed.

Figure 3.17
A tractor-driven fertiliser spreader

- *Placement.* This involves the placement of the fertiliser close to the growing plant. Two methods can be used depending on the type of crop being grown:
  (a) *Ring method.* Here the fertiliser is sprinkled in a ring on the surface of the soil around the growing plant. This is a suitable method for established crops, especially trees (Figure 3.18).
  (b) *Pocket method.* A shallow hole is dug close to the plant and the fertiliser is put in and mixed with the soil. The shallow hole should be made a few inches from the plant and the roots will soon reach the fertiliser.

  Placement is a very effective way of spreading fertiliser. The farmer makes sure that the fertiliser is placed near plant roots, where they can make best use of it.
- *Liquid feeding.* This involves dissolving a fertiliser in water and applying it to the plants. This can be carried out on very large

Figure 3.18 Placement of fertiliser by the ring method (left)

Figure 3.19 Fertiliser can be dissolved in water and applied through a watering can. This is a good method of feeding seedlings in a nursery (right)

farms by machine or may be done on a small scale by hand. This method is very suitable for young seedlings in the nursery (Figure 3.19). Organic manures, if soaked in a barrel of water, make a very good fertiliser for a nursery.

When to apply fertiliser

Farmers may apply fertilisers at different times of the growing season to different crops depending on their needs.

Like humans, when crops are young and growing quickly, they need a good supply of nutrients. For this reason fertiliser is often added to the seed-bed at the time of planting. As the seedlings grow, they have a good supply of food.

Sometimes a crop may need some fertiliser once it is established and growing. Fertiliser may be applied directly to the growing crop; we have already seen that this is called top dressing.

Small amounts of fertiliser may be given to a crop as it grows and as it produces fruit or sets seed. When fertiliser is given to a crop several times in the growing season it is called split dressing.

We will look more closely at the special needs of crops for fertilisers in later books.

ACTIVITY 2

The effect of fertiliser on a crop

This is a class exercise for which you will require:

- a plot of land (4 m × 4 m)
- mixed fertiliser (200 g of 20:10:10)
- rice seed (325 g)
- tools

Your teacher will help organise this activity. This is what you do:

1 Select a small site, approximately 4 m × 4 m, in your school garden. Try to find an area that has not had any manure or fertiliser on it last year.
2 Dig it, take out all the weeds and rake it to a fine tilth.
3 Scatter a small quantity of rice seed all over it and rake the seed well in.
4 Make sure that the plot is kept watered.
5 When the seedlings are 7–8 cm high, divide the plot in half with two sticks and a piece of string. Now scatter the fertiliser that your teacher has given you on half of the plot only. Do not let any fall on the other half. Ensure that water is applied to all the plot to dissolve the fertiliser. Water both sides of the plot to make sure all plants get the same treatment!
6 Every three days inspect the plots and note any difference between the growth and colour of the rice. If one is higher than the other try to measure the difference.
7 Keep a regular record in your notebook of what is happening to your crop. If you can keep your test going so that you can measure the weight of grain produced from each plot, so much the better.

Has the fertiliser made any difference? Why is this?

Some key words and terms in this chapter

Plant nutrients Simple substances taken up by plant roots as a source of food.

Chlorophyll The substance that gives plants their green colour and is responsible for photosynthesis.

Photosynthesis The process in plants in which carbon dioxide from the air, and water from the roots, is turned into sugar.

Stomata The holes, or pores, in the leaves of plants which allow the plant to breathe and to lose moisture.

Transpiration The loss of moisture from the leaves of a plant. It keeps the plant cool and draws up nutrients from the roots.

Seed Seeds are produced after a flower has been pollinated. The seed ensures the next generation of plants.

Germination The sprouting of a seed leading to the development of a new seedling plant.

Soil fertility A measure of the potential of a soil to produce crops.

Fallowing Allowing a plot of land to rest between crops, which lets soil fertility build up.

Slash and burn agriculture A traditional form of farming in Africa. Farmers cut down the bush, burn the trees and grow crops on the area for a number of years before moving on.

Humus Produced when organic matter decomposes in the soil.

Organic fertiliser A manure of plant or animal origin.

Inorganic fertiliser A fertiliser which comes from an industrial source.

N, P and K The three letters normally used to mean nitrogen, phosphorus and potassium, the three main plant nutrients commonly added to the soil in fertilisers.
Straight fertiliser A fertiliser consisting of one material only, supplying only one plant nutrient.
Mixed fertiliser A fertiliser supplying two or more plant nutrients.
Compound fertiliser See mixed fertiliser.
Major plant nutrient One of a group of plant foods required in quite large amounts by plants.
Minor plant nutrient One of a group of plant foods needed by plants in relatively small, but essential, amounts.
Trace element See minor plant nutrient.
Deficiency disease A disease in a plant (or animal) caused by a lack of some nutrient.
Broadcasting Spreading fertiliser evenly all over a piece of land.
Placement Putting fertiliser close to a plant where it will be easily available to the plant.
Leaching The loss of nutrients from the soil in drainage water.
Prills Fertiliser which is made in very even, tiny, free-flowing balls.
Ratio The relative proportion of one thing to another, by weight, percentage, length etc.

Exercises

Multiple choice questions

Write the correct answers in your exercise book.

1 Which part of the plant holds it firmly in the soil?
  a stem c roots
  b flowers d branches

2 The part of the plant where food is manufactured is the
  a trunk c roots
  b leaves d stalks

3 The green substance in the leaves of plants is called
  a chlorophyll c flowers
  b stomata d fruits

4 The tiny holes found underneath the leaves are
  a fruits c stomata
  b flowers d branches

5 Which plant nutrient is given the letter K?
  a potassium c phosphorus
  b sodium d iron

6 An example of an organic fertiliser is
  a urea
  b potash
  c farmyard manure
  d ammonium nitrate

7 Which one of these is an aid to soil fertility?
  a leaching
  b brushing the land
  c drought
  d green manure

8 A method of applying fertiliser is
  a planting a seed
  b transplanting
  c ring method
  d fallowing

9 A farmer has 250 kg of a 20:10:10 fertiliser to apply to one hectare. How many kg of plant nutrients will he apply?
  a 50:25:25
  b 100:50:50
  c 150:75:75
  d 200:100:100

Matching items

10 Match the definition with the most appropriate answer from the list below:

a Spreading fertiliser all over a plot of land.
b Putting fertiliser in a hole close to the plant's roots.
c Applying fertiliser to a growing crop.
d Applying fertiliser several times during the growing season.
e Digging in a growing crop to supply organic matter to the soil.

top dressing	**split dressing**
broadcasting	**pocket method**
green manuring	

Missing words

11 Complete the following sentences by selecting the most appropriate words from the list below. Write out the sentence in your exercise book.

Growing crops need water and ...... from the soil. The ...... are able to select what is required from the soil and pass it up through the ...... to the leaves of the plant. The leaves of the plant are able to make sugar from water and .... from the air; this process is called ...... The sugars are combined with the minerals from the soil to make all the plant's needs; these are passed around the plant through a series of ......

vessels	**photosynthesis**
carbon dioxide	**stem**
nutrients	**roots**

Crossword puzzle

12 Copy the diagram into your exercise book and complete the following clues.

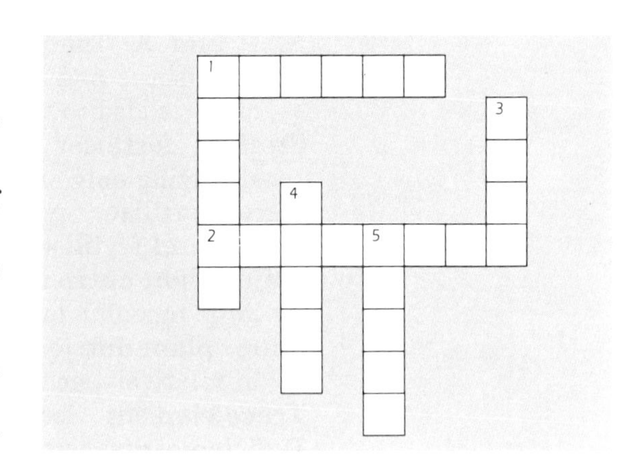

Across
1 Describes an inorganic fertiliser which is ground up into a fine dust (6)
2 The loss of plant nutrients from the soil through drainage water (8)

Down
1 Describes the small shiny regular balls which are used to make up some inorganic fertilisers (6)
3 A method of placement of fertiliser around the plant (4)
4 This is lost from the leaves during transpiration (5)
5 This is formed when organic matter rots down in the soil (5)

Points for discussion

13 Farmers grow a particular crop because it has some part or parts which are useful. For example, coffee is grown for its useful seeds. Make a list of all the crops grown in your area and identify the useful parts of the plants. Have you been able to think of crops which exploit every part of the plant?

14 How fertile is the soil in your area? What do farmers do to maintain soil fertility? Do they use compost or inorganic fertilisers? Which farmers use which materials? Why is this?

15 Where are farmers able to buy inorganic fertiliser? What types are available? Who uses it? What is it used for? How is it applied?

CHAPTER

4

Simple farm and garden tools

Introduction

In the previous chapters we learnt about the soil, the fertilisers that farmers add to it, and how farmers make soil conditions right for the growing of good crops. We have not yet thought about the tools that are used when farmers work in the fields. In this chapter we will learn about hand tools; in later books we will look at the work that machines like tractors can do.

The origins of hand tools

The hand tools we use today have been developed from those used by early people. The early tools made of rough sticks, stone and animal horns have been replaced by those made of shaped wood and sharpened steel.

Hand tools play a very important part in the agriculture of Sierra Leone, where many farmers are not able to use machines. Hand tools make work much easier and allow farmers to create the soil conditions that are needed to grow crops. Think how difficult it would be to work in the fields without any tools; you would only have your bare hands!

The hand tools we use today have many advantages. The steel does not wear away quickly and can be sharpened easily. Steel tools can be made or repaired by local blacksmiths. The wood used for handles is cheap and easy to replace if it breaks, as it can be cut locally and carved into the required shape. Wood is long lasting, especially if stored in a dry place. Wooden handles are able to absorb shocks when the tools hit the soil; this means that you do not feel vibrations in your arms when using something like a hoe!

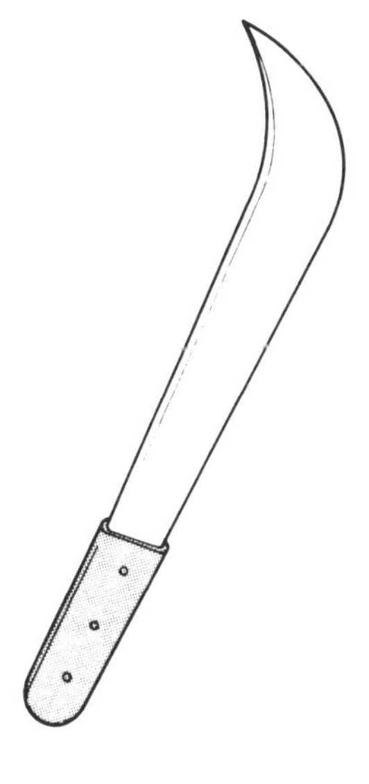

Figure 4.1
Cutlass

Cutlass

As you can see from Figure 4.1 a cutlass is like a large knife; it has a steel blade about one metre long, which is sharpened on one side; it also has a wooden handle made up of two parts, which are riveted together. The cutlass is a very adaptable tool and has many uses.

The main use of a cutlass is for cutting. Before you can start cultivating a garden it is necessary to clear all the vegetation; the cutlass can cut through green plants, thin woody stems or small branches and is ideal for this purpose. Cutlasses are often used for many jobs for which they were not designed! A cutlass is an ideal tool for working small areas of soil, for digging small holes for transplanting seedlings, for weeding, for opening up the surface of the soil or for harvesting fruit, such as bananas. What jobs have you seen being done with a cutlass?

For keeping the grass or weeds short around a house, many people use a form of cutlass known as a slasher. This consists of a long thin piece of steel, one end of which is bound tightly to a short wooden handle. One edge of the steel is sharpened. The blade may be half a metre long. Disused straps from big packing cases are sometimes made into slashers and are very effective.

A cutlass or slasher will need to be kept very sharp. When using one, you should take care to cut away from yourself and others. Be very careful not to cut yourself when sharpening a cutlass; always sharpen it away from yourself.

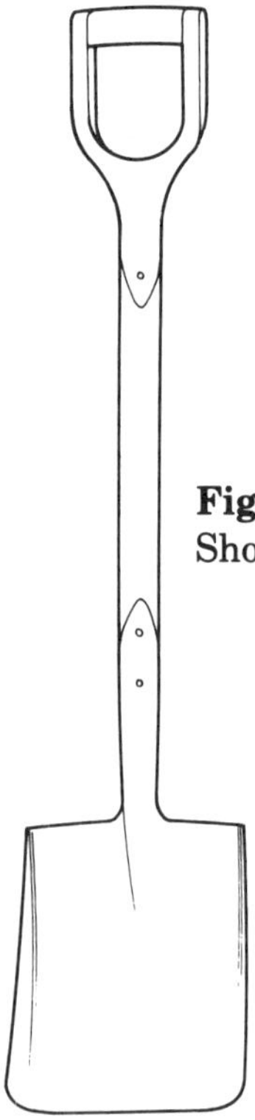

Figure 4.2
Shovel

Shovel

This consists of a wooden handle attached to a broad, curved, metal blade which can have a straight or rounded end; the blade can also be heart-shaped. A shovel is used to move materials. For example, a shovel can be used to load compost from ground level into a wheelbarrow. Shovels may be used for making garden paths, for levelling the ground and for mixing mortar. The blade shape makes a shovel unsuitable for digging (Figure 4.2).

Spade

The spade has a wooden handle attached to a broad, rectangular metal blade (Figure 4.3). The edge of the blade is sharp so that it can easily be driven into the soil. A spade is used by pushing the blade into the soil, using the foot to press down on top of the blade until it is deep into the soil, and then turning over the soil (Figure 4.4).

Spades are mainly for seed-bed construction and for straightening the edges of seed-beds. They are also used for digging water channels and ditches, and for uprooting rice seedlings for transplanting.

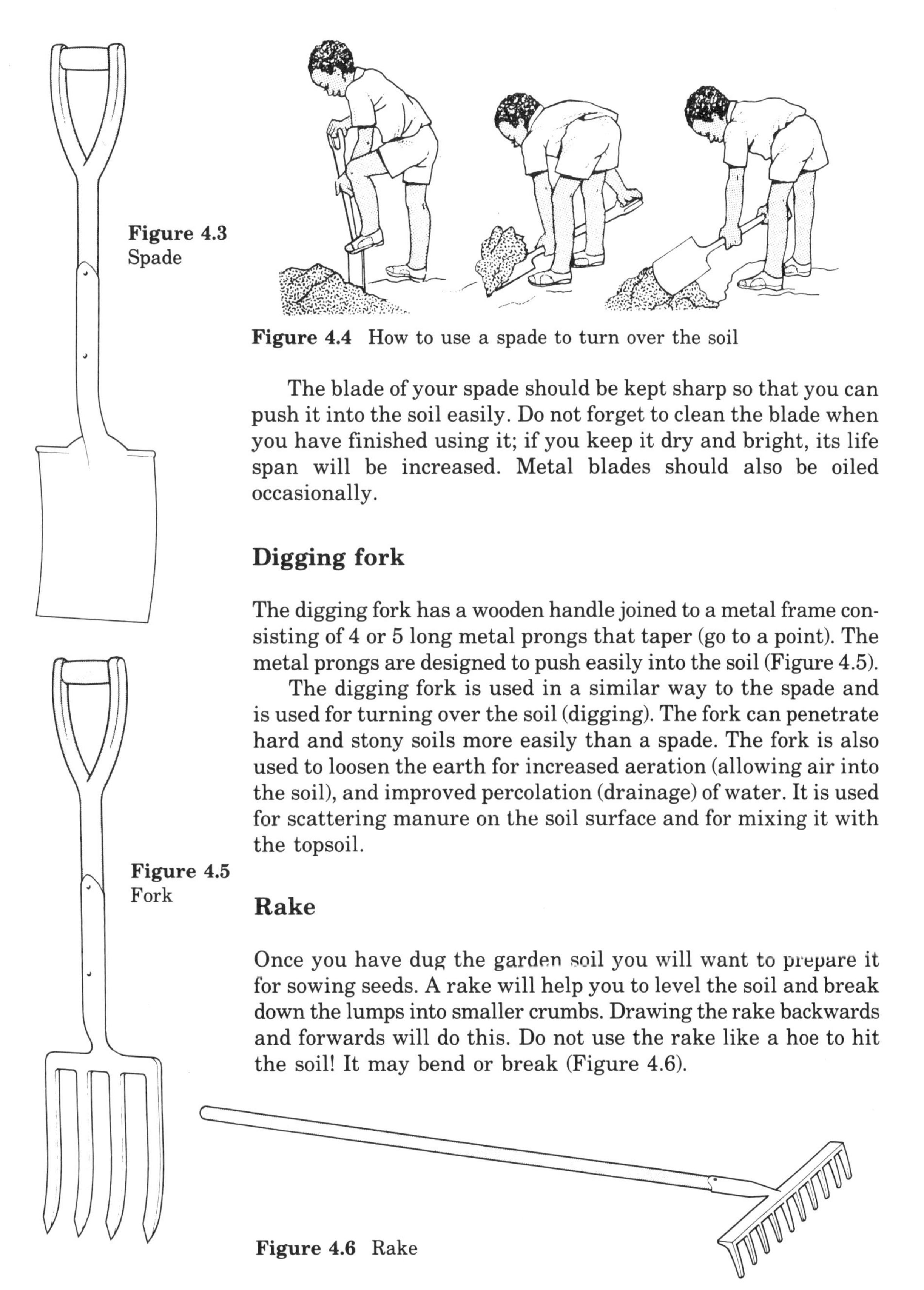

Figure 4.3 Spade

Figure 4.4 How to use a spade to turn over the soil

The blade of your spade should be kept sharp so that you can push it into the soil easily. Do not forget to clean the blade when you have finished using it; if you keep it dry and bright, its life span will be increased. Metal blades should also be oiled occasionally.

Digging fork

The digging fork has a wooden handle joined to a metal frame consisting of 4 or 5 long metal prongs that taper (go to a point). The metal prongs are designed to push easily into the soil (Figure 4.5).

The digging fork is used in a similar way to the spade and is used for turning over the soil (digging). The fork can penetrate hard and stony soils more easily than a spade. The fork is also used to loosen the earth for increased aeration (allowing air into the soil), and improved percolation (drainage) of water. It is used for scattering manure on the soil surface and for mixing it with the topsoil.

Figure 4.5 Fork

Rake

Once you have dug the garden soil you will want to prepare it for sowing seeds. A rake will help you to level the soil and break down the lumps into smaller crumbs. Drawing the rake backwards and forwards will do this. Do not use the rake like a hoe to hit the soil! It may bend or break (Figure 4.6).

Figure 4.6 Rake

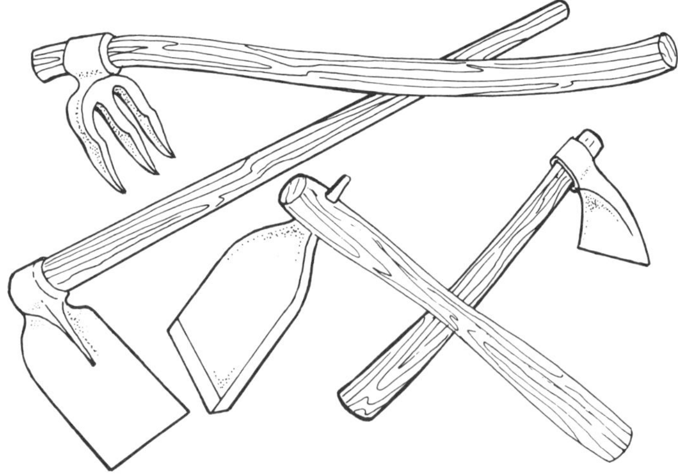

Figure 4.7
Different types of hoe

Hoe

Hoes vary in shape from country to country, but they all have a common use, to break up the topsoil. In doing this, they can remove weeds and open the soil to let in air. The depth a hoe will go into the soil depends on the length of the handle and blade. Several hoes are illustrated in Figure 4.7.

The most useful hoe for your garden will have a short wooden or steel handle with a blade that should not be more than about 10 cm long. The blade should be kept sharp so you can cut weeds. To use a short-handled hoe, you have to bend near to the ground and you can weed accurately without cutting the vegetables.

West African hoe

The West African hoe normally has a short wooden handle and a narrow blade sharpened at the end. The blades are made by local blacksmiths or may be imported. Most hoe blades are held on to the shaft by a narrow pointed spike that is forced through a hole in the bottom of the wooden handle. This hoe is used for shallow digging of the soil, weeding, making mounds and seed-beds. It is also used for making shallow holes for crop planting, and for harvesting some root crops.

With a long-handled hoe of this type you can hit the soil from a standing position; as the blade is not wide it will go deeper into the soil. It is used on small farms for breaking up lumps of soil and for turning over land that has been ploughed.

West Indian hoe

The West Indian hoe has a long wooden handle which needs less energy to use. It has a metal blade that is broad and oval. It has a number of uses: it can be used to cut up roots, stones can be removed from seed-beds, it can turn over clods of soil and it can bury weeds. This is possible as the blade is large and can penetrate quite deeply into the soil. This hoe is also used for digging soil in swampy areas.

Ridging hoe

This is a tool with a concave blade (hollowed out) and a bent-back handle. The hoe scoops the broken soil, turning it over to make one side of a ridge and, at the same time, burying the weeds. The ridge is completed by moving along the other side and turning the soil back against the first half of the work. Farmers in swampy areas use a long-handled hoe which has a straight blade of wood tipped with metal to make ridges. Neither of these hoes is used in the vegetable garden.

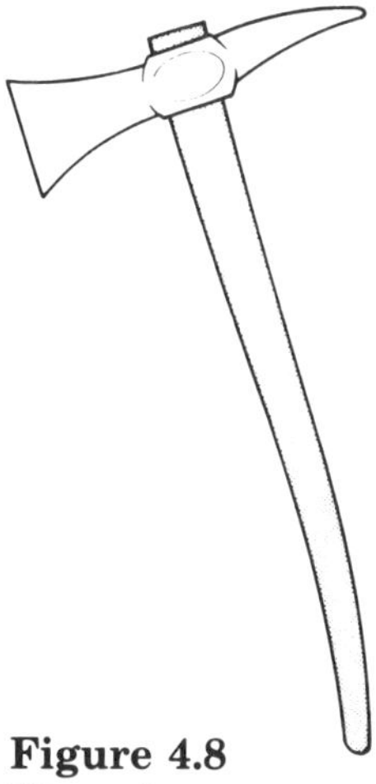

Mattock

The mattock is rather like a short-handled hoe with a strong handle and a head consisting of two wide blades. One blade is straight with a sharp edge and the other blade is curved and is used for levering out roots (Figure 4.8). This is a strong tool for use in difficult soils.

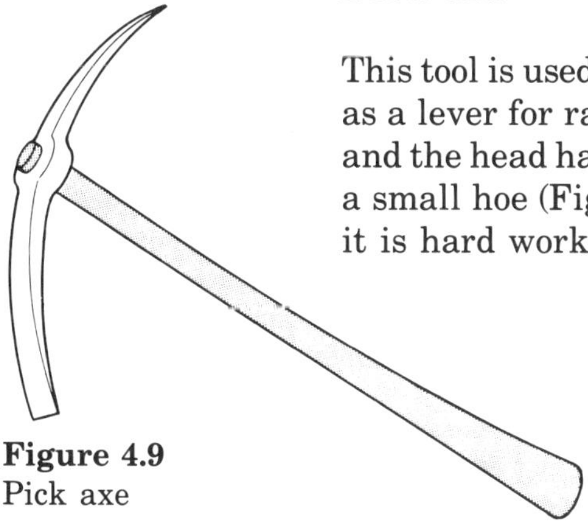

Figure 4.8
Mattock

Pick axe

This tool is used for digging holes (especially in stony ground) and as a lever for raising big stones. The handle is short and strong and the head has two ends, one of which is pointed, the other like a small hoe (Figure 4.9). As it is heavy you will find that using it is hard work!

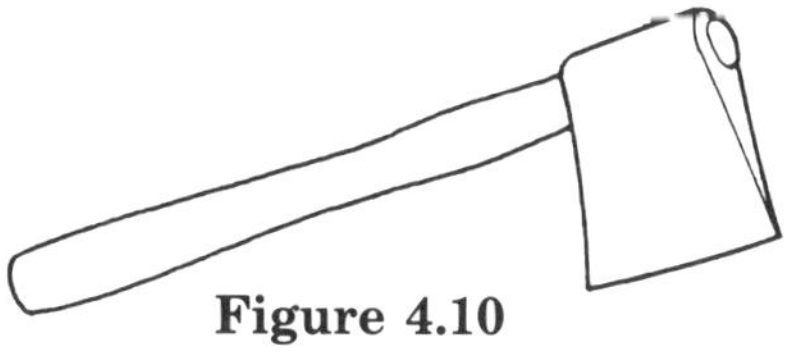

Figure 4.9
Pick axe

Figure 4.10
Axe

Axe

The axe is used for chopping wood. The heavy one-sided head is fitted with either a straight or curved handle. The head is held in place by a wedge driven into a slot in the top the wooden handle

(Figure 4.10). To get the best use from an axe, the cutting edge must be kept very sharp and resharpened regularly when in use. If the head becomes loose, stand the axe in a puddle or bucket of water overnight; the wood will swell and grip the head more tightly. You should be taught how to use an axe before trying your skill. It can cause accidents.

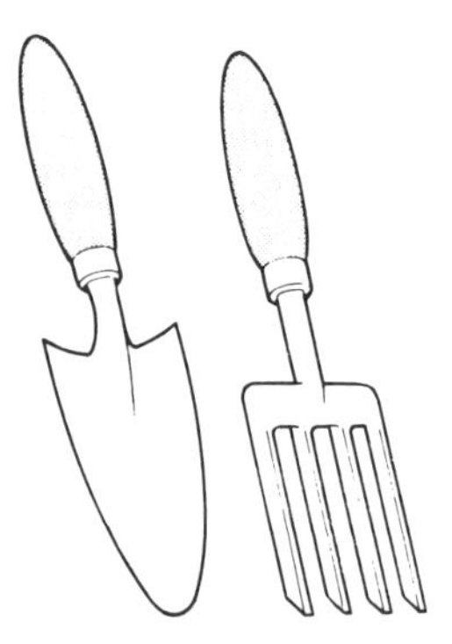

Figure 4.11
Hand trowel and hand fork

Hand trowel

The hand trowel is a small hand tool consisting of a short wooden handle and a metal blade that looks like a large spoon (Figure 4.11). It is an excellent tool for transplanting seedlings and for making small planting holes. The curved blade makes it possible for a small ball of earth to be carried with the seedling. Transplanting means moving a plant from one growing place to a new position. Usually we move a seedling or young plant from a nursery to the place where it will grow and mature.

Hand fork

This small tool has a short wooden or metal handle and a metal blade with prongs (Figure 4.11). The fork could have 3, 4 or 5 prongs that taper at the end so that the fork can easily penetrate the soil. It is used for light weeding and for loosening the soil surface around seedlings; this allows roots to penetrate the soil. The pegs of groundnuts must grow down into the soil and benefit if the soil is loosened with a hand fork. Hand forking increases aeration and percolation of water into the soil. It may also be used for mixing manure or fertiliser into the soil over a small area.

Hand cultivator

This has a metal blade and metal handle. The metal blade is pronged and bent at an angle to the handle. It is used for light tilling of the soil, light weeding, and for mixing the topsoil with manure or fertiliser.

Watering can

This tool is a most useful piece of equipment because it helps you to provide each plant with water, accurately and without waste. It holds enough water for you to carry comfortably to the plants (Figure 4.12). With a rose on the end of the spout, water can be

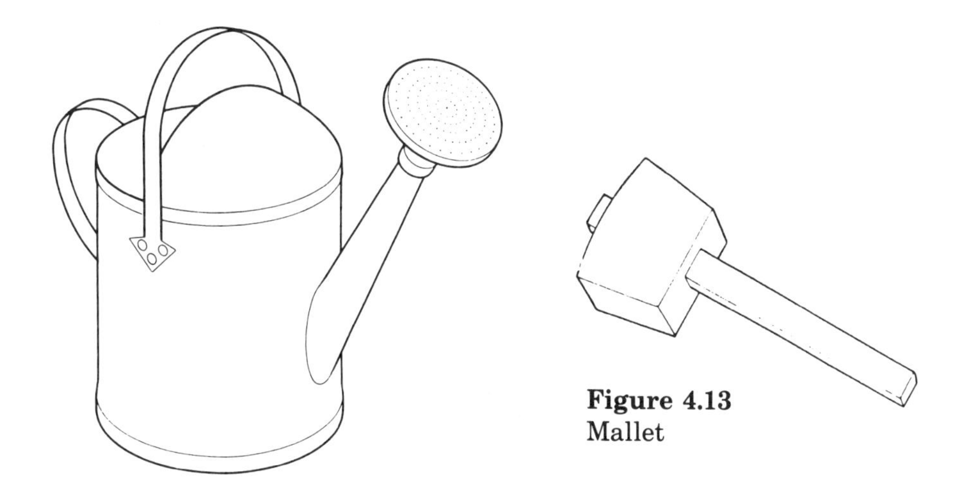

Figure 4.12
Watering can

Figure 4.13
Mallet

applied gently without washing the soil away or damaging plants. You may use the can without the rose; by placing your finger over the end of the spout and pouring slowly it is easy to direct the amount of water you want on to the soil around a plant.

Sieve

Sieves are used for grading (sorting out) small quantities of soil which are often needed for filling seed boxes. Sieves are made with either a coarse or a fine mesh. By filling a sieve with soil, picking it up and moving it quickly from side to side, lumps of soil and stones that are bigger than the mesh size will not fall through while the fine particles of soil will. This fine material will make excellent soil for seed boxes.

Mallet

This tool, with its wooden head and handle, is used to drive in pegs or fence posts (Figure 4.13).

Measuring tape

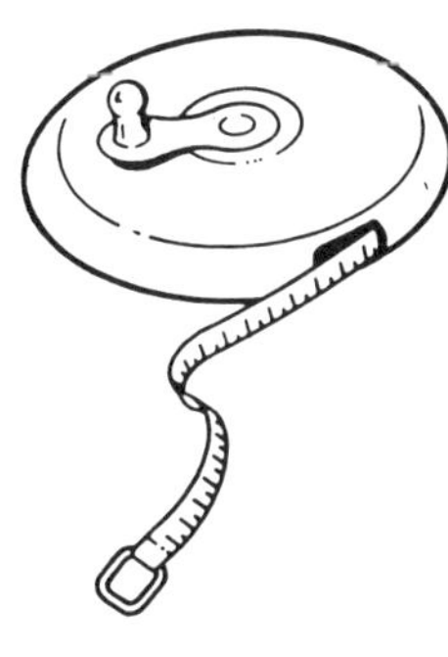

Figure 4.14
Measuring tape

The measuring tape is a useful aid to farmers for such jobs as tree planting, row spacing or constructing buildings. The tape may be made of plastic or metal and is usually wound up in a carrying case (Figure 4.14). Tapes are available in an assortment of lengths such as 30 m and 100 m. Tapes are either in metric (metres, centimetres) or imperial units (yards, feet, inches), or both.

ACTIVITY 1

Using a tape measure

This is an exercise for the whole class and one that will require the help of your teacher. For this exercise you will need a measuring tape, 25 wooden pegs or sticks and an open piece of flat ground 20 m × 20 m (Figure 4.15).

You are to imagine that we are going to plant out fruit trees in a square with the trees spaced by 3 m in each direction. This is what we do.

1 Using your tape measure, construct a 3–4–5 right-angled triangle. Lay out the triangle close to the edge of the plot; mark the points of the triangle with three pegs. The peg at the right angle marks the first corner of your plot. (Your teacher will explain in more detail about the use of a 3–4–5 triangle to make a right angle.)
2 Measure out 12 m along the two shorter sides of the triangle you have just created and mark the spots with pegs; these will mark the next two corners of your plot.
3 Repeat this operation with the 3–4–5 triangle at the two corners you have marked to complete the square. You can check that your plot is square by measuring diagonally across the plot; the measurements should be the same.
4 Complete the grid pattern by marking out lines at 3 m intervals. Knock in a peg at every point. Your 'orchard' should now be complete with 25 trees planted!

Figure 4.15
Using a tape measure to mark out a plot

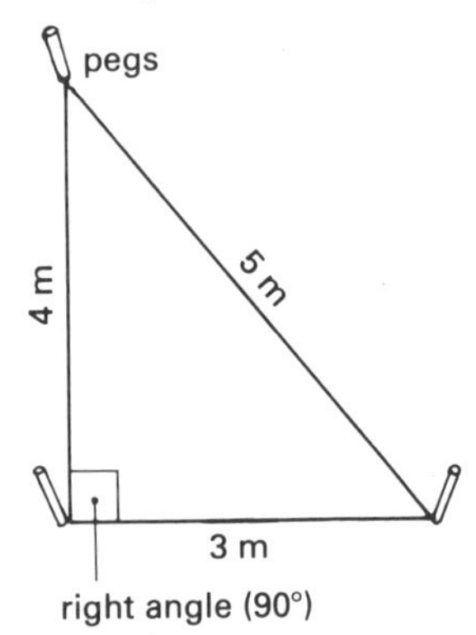

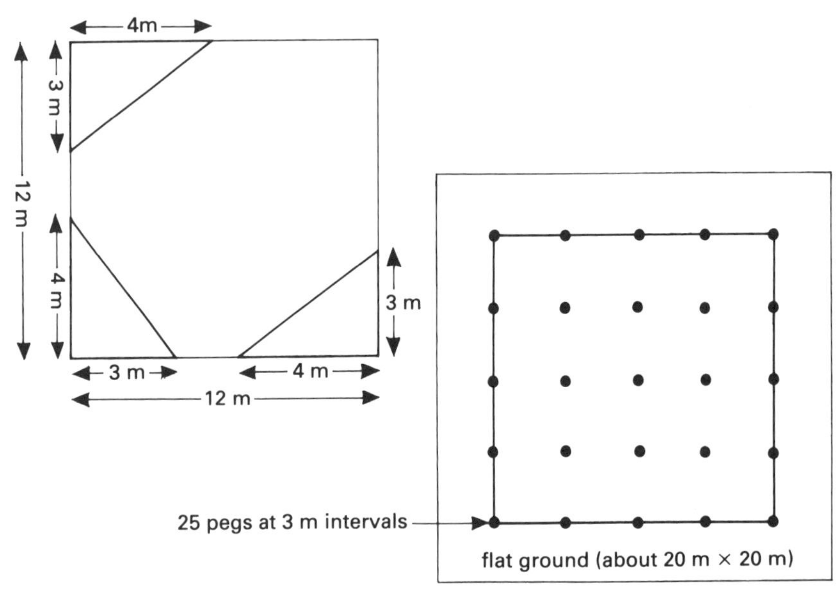

If the ground is not very even you can have difficulty getting measurements exact. This is a problem farmers face when planting out large numbers of trees on uneven ground; some alignment by eye may be necessary!

5 If your orchard has been planted successfully you should be able to see straight lines up and down and diagonally across the plot! Did you plant your 'trees' properly?

This exercise is also useful if you want to do some building.

Garden line

When transplanting, sowing seeds or marking out plots, a garden line is essential.

ACTIVITY 2

Making a garden line

It is very easy to make your own garden line! You will need two sticks about 40 cm long, a length of string (perhaps up to 12 m), some coloured string or strips of cloth and a ruler for measuring (Figure 4.16).

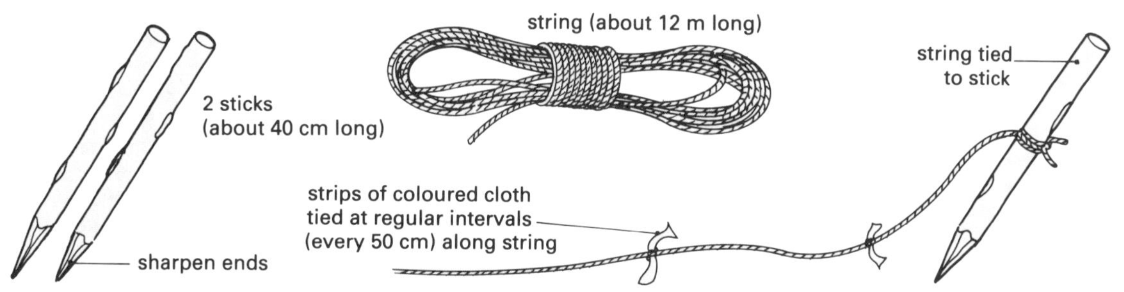

Figure 4.16 Making a garden line

This is what you do:

1 Sharpen one end of the sticks so that they can be pushed easily into the soil.
2 Tie the string to the centres of the two sticks.
3 Take your line and tie small pieces of coloured string or cloth at regular intervals along the line. You will have to use a ruler for this. Choose a spacing which will suit the crops you will be planting out, such as every 50 cm.

Your garden line is a cheap and accurate tape measure and will ensure that each plant is the same distance apart. After use, roll the string on to one stick only, keeping the other end tied to the other stick.

Knapsack sprayer

The knapsack sprayer is used for spraying chemicals on to crops. Farmers may want to spray insecticides, weedicides or fungicides (to kill insects, weeds or fungi respectively, which may be attacking the crops).

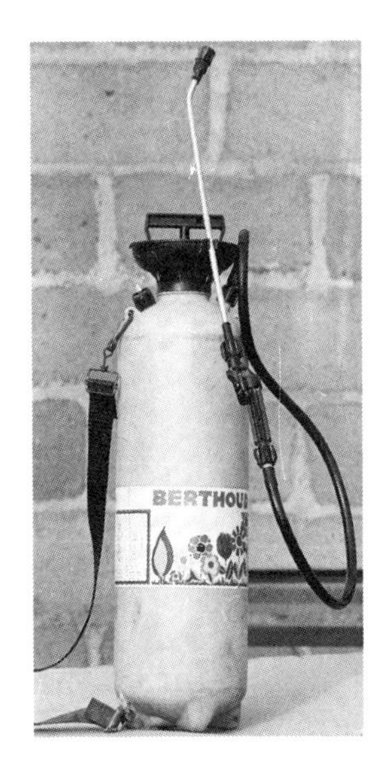

Figure 4.17
Knapsack sprayer

The knapsack sprayer consists of a metal or plastic tank which holds the spray. It has a long tube with a nozzle at the end from which liquid is forced out as a fine spray (Figure 4.17). It is normally carried on the back of the user. The operator has to pump the sprayer using the handle to force the spray out.

If you are using spray chemicals you must be very sure to follow all the instructions that come with the chemical. Some chemicals are **very** dangerous!

Wheelbarrow

The wheelbarrow consists of an open box with a wheel at one end and a pair of handles at the other (Figure 4.18). With it you can move materials about easily. If the wheel has a solid rubber tyre it will last a long time.

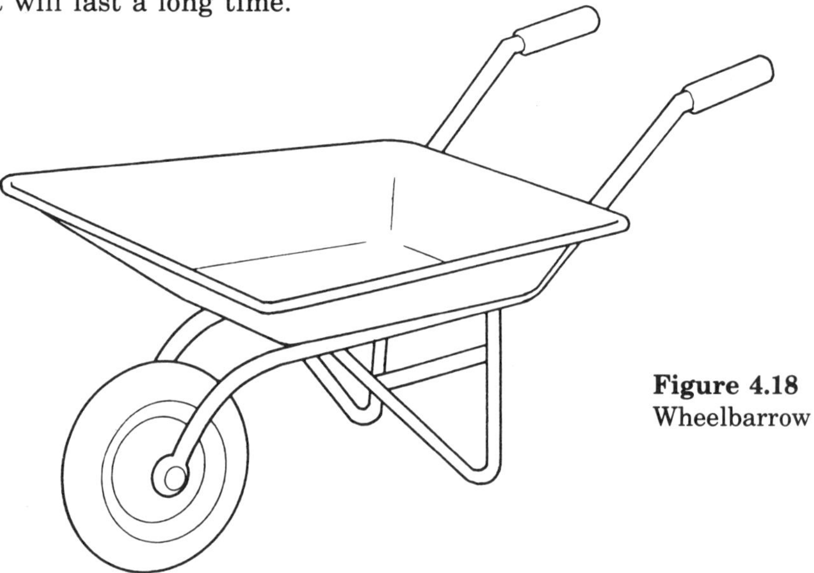

Figure 4.18
Wheelbarrow

The secret of using a wheelbarrow is to load it with the weight as near to the wheel as possible. In this way your arms will have less weight to lift. Keep the wheel axle well oiled and never let the barrow stand for long periods with water in it. When not in use, it should be stood up on the wheel end with the handles resting on a wall.

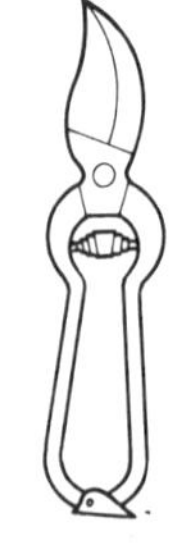

Figure 4.19
Secateurs

Secateurs

Secateurs are used for the clipping and cutting of hedges or pruning thin woody plants or shrubs. They can also be used for the cutting of budwood in budding operations. Secateurs have bare metal handles which are sometimes covered with rubber and are connected to two short metal blades (Figure 4.19). A metal spring is usually fitted between the handles. Secateurs cut using the lever principle whereby the handles are squeezed together over a long distance to make the blades move very powerfully through a short distance. Secateurs are sharpened like a pair of scissors.

Garden shears

Garden shears work like a large pair of scissors and, like secateurs, use the lever principle. Shears have wooden or rubber handles with long metal blades that have sharp cutting edges (Figure 4.20). They are used for cutting hedges, for trimming plants with thin woody stems and for cutting grass.

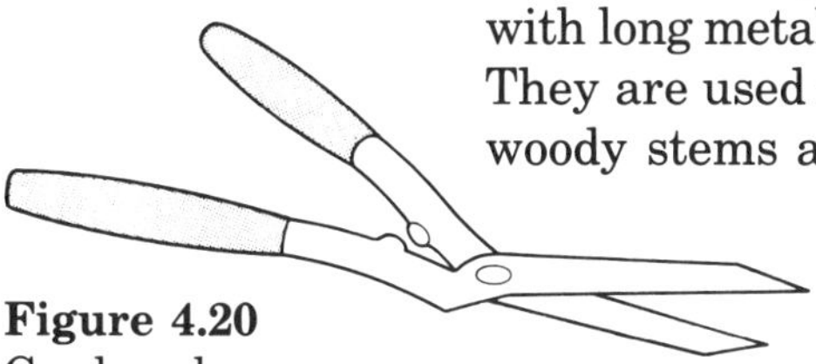

Figure 4.20
Garden shears

Sickle

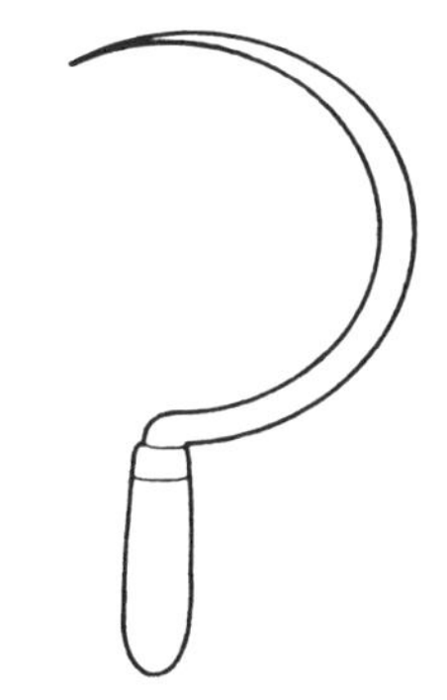

Figure 4.21
Sickle

This tool has a curved metal blade shaped like a half moon attached to a wooden handle (Figure 4.21). It is used for harvesting rice, grasses and fruits.

Long-handled sickle or 'go-to-hell'

This tool has a normal sickle blade which is mounted on a long wooden shaft. It is used for harvesting fruits found on tall trees, such as cacao.

Pruning saw

The pruning saw has a wooden handle and a metal blade that has saw teeth on both edges (Figure 4.22). One set of teeth is large and coarse for cutting thick branches; the other set of teeth is smaller and finer for use on thin branches or woody stems. It is used for pruning.

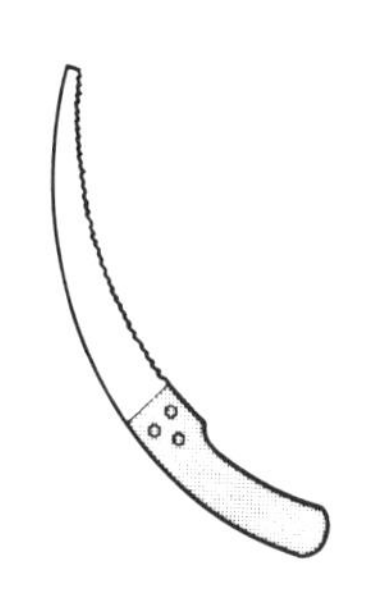

Figure 4.22
Pruning saw

Another type of pruning saw has a long, curved, saw-toothed blade. The teeth cut as the saw is pulled downwards. It is used for cutting off branches from trees and, in particular, for harvesting oil palm fruit.

Pruning involves cutting off dead or diseased branches of trees, cutting off living branches to check unwanted growth, cutting away branches to allow more light to enter the tree, and helping to maintain the desired shape of a tree.

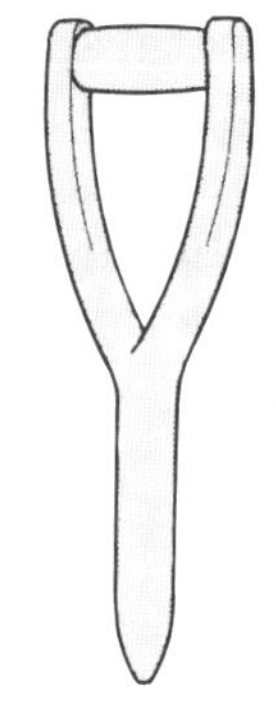

Figure 4.23
Dibber

Dibber

This is a very simple and useful tool! A dibber is used to make holes in the ground for planting or transplanting seeds, roots, bulbs or tubers. It consists of a pointed wooden shaft attached to a handle. The broken shaft of a spade can make an excellent dibber (Figure 4.23).

Scales

Many different types of scales are available. A common design consists of a pan (a flat plate or bowl) which holds the material to be weighed. This is placed on the body of the scales, which is made from metal or plastic. The weight of the material in the pan is shown on a graduated scale by a pointer. The scale is graduated in metric (kilograms, grams) or imperial (pounds, ounces) units.

ACTIVITY 3

Identifying gardening activities and tools

In Figure 4.24 try to identify all the activities taking place, and all the tools being used.

Figure 4.24

Looking after tools

General care

If a farmer's tools are to be reliable and last a long time, they must be properly looked after. Tools are just too expensive to neglect!

We cannot allow the metal parts of tools to rust as the tools will not last long and will be difficult to work with. In order to keep the tools bright and shiny, the metal parts must be lightly oiled after use. Any type of oil or grease will do and it can be put on with a cloth. If you have been digging, when you have finished work, you should scrape off soil and mud from spades and forks before oiling them. You can use a small flat piece of wood for this.

Never leave tools out in the rain as the metal will rust, the wood will rot and they will eventually be ruined.

Sharpening

If tools are to work effectively they must be sharp; a sharp tool will be easier to use and will need less effort. It is best to sharpen tools when you have finished work; they are then stored away safely and are immediately ready for you next time!

It is easy for you to sharpen your own axes, cutlasses, hoes and spades. It will not be possible for you to sharpen secateurs, shears and pruning saws; these tools are more difficult to sharpen and need special tools.

This is how you can sharpen some of your simple tools. If you are lucky enough to own a carborundum stone or a fine file you will find they will sharpen most tools.

If you do not have a carborundum, you can use a stone from the ground! Select a hard stone from the ground; your teacher will tell you which is the best. Rub the stone along the blade away from you with the edge of the blade pointing away. Test for sharpness by drawing a leaf along the cutting edge or use a piece of paper.

It is possible to sharpen thin blades, such as knives, with a hammer. This method is not very good and should only be used if no other method is possible! Lay the blade side down on a hard flat surface like a closed vice or an anvil (a large block of metal used by a blacksmith). With the rounded end of a hammer, tap the edge of the blade gently to flatten and sharpen it. The cutting edge will tend to be uneven along its length and the process is rather slow.

Do not forget that any tool with a sharp edge can be dangerous. Do not play with sharp tools or use them incorrectly.

Replacing broken parts

You must expect wooden handles to break through normal wear and tear. Replacing them is not difficult. You should use only weathered wood (wood that has been stored in a dry place for at least six months). If you use a piece of wood cut directly from a tree, the wood will still have sap in it and will dry out and shrink. It will no longer fit the tool closely and the head will become loose, making the tool dangerous or useless.

The broken pieces of a handle should be laid out together so that the new one can be made the same size and shape.

The fitting of new handles to hammers and axes needs special attention. The handles of hammers and axes have a saw cut made across the head end. After fitting the head, a wedge of hard wood is driven into the saw cut making a really solid joint; excess wood is then cut from the wedge.

Forks and spades with metal straps extending from the blade up the handle are not so easy to repair. You will either need to cut the rivets holding the broken handle between the straps using a hacksaw, or file the tops off the rivets which will allow you to knock them through with a hammer. When you have cut a new handle, push it well down between the straps and drill holes for the new rivets. Large nails can be used if no rivets are available. Make sure that the ends of the nails are cut off and rounded over. If you just bend the nails down, you may find they catch your hands when you are using the spade or fork.

ACTIVITY 3

Here are a number of activities you can do on your own or as a class.

Visit to a local blacksmith
Visit your local blacksmith. Ask him what he does and how he does it. What tools does he repair for farmers? How does he do this? Draw a picture of the tools he uses.

Visit to a local tool store
Visit your local tool store or agricultural merchant. What agricultural tools are for sale? Make a list of them and write down how much they cost. Which tools are made locally and which tools are imported?

Visit to a farmer
Visit a local farmer. Ask him what tools he owns. Write down a list of them. What tools does he repair himself? What wood does he use and where does he get it? Ask him what tools he is missing and which he would like most. Why would he want these particular tools?

Some key words and terms in this chapter

Aeration Allowing air to enter the soil.
Percolation The passage of water down through the soil.
Transplanting Moving of plants from one place to another. Usually this applies to moving seedlings from a nursery bed to the place where they will grow and mature.
Seasoned wood Wood that has been cut and allowed to dry out for a period of time before use.
Metric The standard of measurement that has now been accepted in nearly every country of the world. Measurements are based on units of 10. The kilogram, kilometre and litre are examples of metric measurements.
Imperial The old standard of measurement used in countries that were part of the British Empire. These units of measurement are now going out of use. The pound, the mile and the gallon are imperial measurements.
Insecticide A material for killing insects.
Weedicide A material for killing weeds. It may also be called a herbicide.
Fungicide A material for killing fungi. A fungus is a very small simple plant. Some fungi can attack crops.
Lever A mechanical aid to make work easier.
Pruning The selective cutting of branches from a tree or shrub, or the removal of diseased branches.

Exercises

Word puzzles

1 Copy the pattern of circles in Figure 4.25 into your exercise book and see if you can complete the puzzle.

Figure 4.25

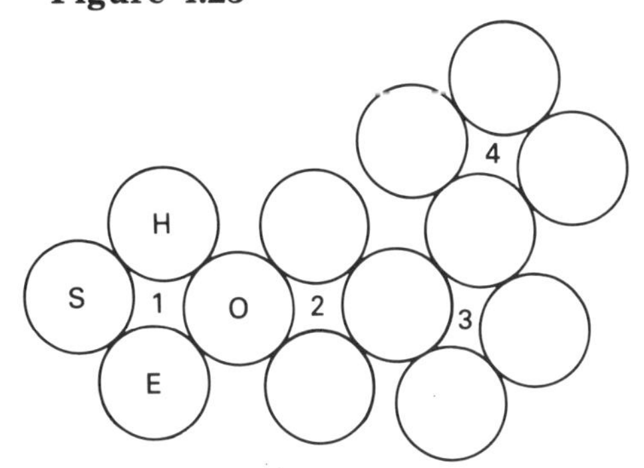

Around each number there are four circles. Fill in the answers to the following questions in those circles. You will find that certain letters are linked. The first one has been done for you as an example.

1 These tools are used for weeding.
2 You can dig with this in your garden.
3 This one helps you to prepare a smooth seed-bed.
4 This will help you to mark out plots in the garden.

Missing words

2 Find the missing word or words from the list below. Write the answer in your exercise book.

a A . . . . is the best tool to use for digging small holes in the garden for seedlings.

b The ...... on a ...... allows water to fall gently on to young plants.
c Digging holes in stony or dry soil is easier if you use a ....
d When loading a ...... with earth, the load should be placed at the ....
e After working in the garden, a good farmer ...... and ...... his spade.
f It is ...... to play games with ...... tools.

hand trowel	**sharp**
dangerous	**fork**
rose	**wheelbarrow**
cleans	**watering can**
front	**oils**

Crossword puzzle

3 Copy the grid into your exercise book and see if you can complete the puzzle.

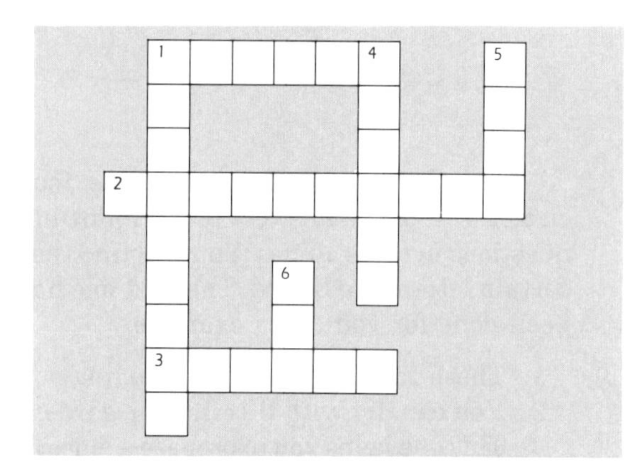

Across
1 These tools are used for digging (6)
2 Used by farmers for spacing out plants (6,4)
3 Used to hold the heads of spades and forks on to the shaft (6)

Down
1 Used for small-scale clipping and pruning (9)
4 These are used to weigh farm produce (6)
5 Used for levelling the soil and making crumbs (4)
6 Used to chop wood (3)

Multiple choice questions

Write down the correct answers in your exercise book.

4 The simple farm tool normally used to level surfaces and to break large soil crumbs into smaller ones is a
a rake
b shovel
c trowel
d spade

5 The tool that should be used for watering vegetable beds is a
a rake
b hand fork
c watering can
d knapsack sprayer

6 Which of the following procedures will ensure the long and effective life of farm tools?
a wash metal blades, dry and oil them
b using them for the wrong job
c allow rust to form on the blade
d leave them in the open

7 The removal of weeds in a vegetable bed is best done by using a
a hand fork
b hand trowel
c sickle
d spade

8 The curved-bladed knife that is used for harvesting rice is the
a shears
b sickle
c secateurs
d pruning saw

9 Which of the following tools is used for pruning?
a hand fork
b secateurs
c sprayer
d mallet

10 Which of the following tools is used for transplanting vegetable seedlings?
a hand trowel
b spade
c hand fork
d hand cultivator

11 Which tool can be used for soil cultivation?
  a sieve
  b axe
  c cutlass
  d rake

12 Which of the following tools is used in seed-bed construction?
  a measuring tape
  b pick-axe
  c pruning saw
  d sprayer

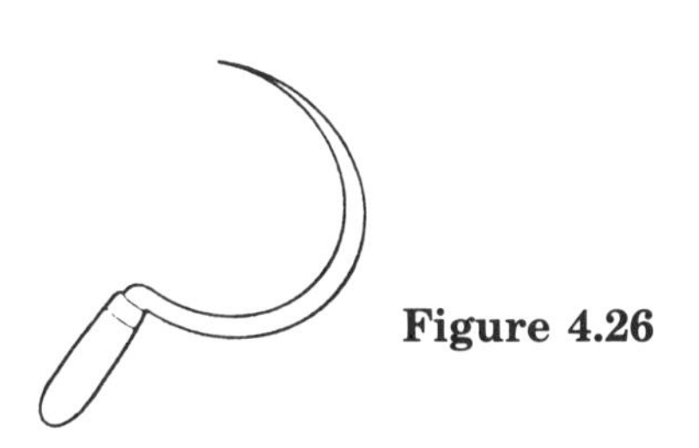

Figure 4.26

Use Figure 4.26 to answer questions 13 and 14

13 The diagram in Figure 4.26 represents
  a secateurs
  b a dibber
  c a sickle
  d a budding knife

14 Figure 4.26 is a tool that is primarily used for
  a harvesting rice
  b harvesting cassava
  c harvesting oil palm
  d harvesting orange

15 Which of the following tools is appropriate for harvesting fruits found on tall trees?
  a sickle
  b 'go-to-hell'
  c pruning saw
  d rake

16 Figure 4.27 represents a
  a hoe
  b garden fork
  c rake
  d 'go-to-hell'

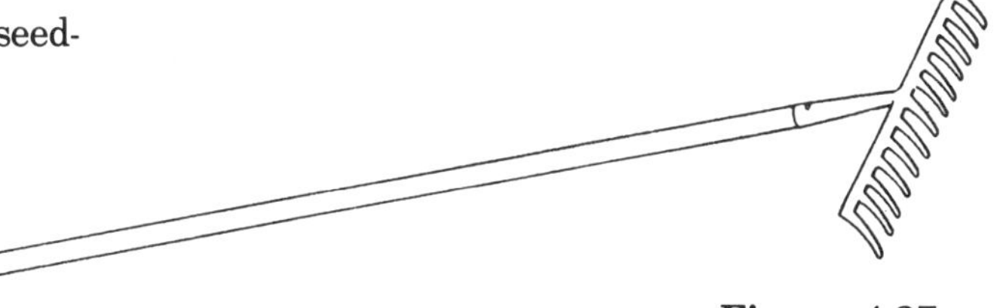

Figure 4.27

Points for discussion

17 What tools do farmers use in your area? Where do they get them? What tools are made locally and what tools are brought in from outside the district?

18 What woods do people use for making tool handles? Are some better than others? Where do these woods grow?

19 How do you think the tools the farmers use are suitable for the soils in your area? Which hoes do they use? Why is this?

CHAPTER 5

Sources of power in agriculture

Introduction

In the previous chapter we looked at the range of hand tools that farmers use and the particular tasks the tools were designed to do. With these hand tools the farmers have to rely only on their own strength to get jobs done. In this chapter we will see that farmers can be much more productive if they can use sources of power other than their own! However, we will find that not all these sources of power are freely available to our farmers in Sierra Leone.

The limitations of human power in agriculture

On visiting farms you will have seen the owners using hand tools to carry out a wide range of tasks. You will probably understand that working with hand tools is always very hard, very slow and very tedious! Have you ever spent a full day digging in the fields?

Have you ever stopped to think about the farmer working with these hand tools. What limits the amount of work that can be done in a day? What could be done to increase it? If the farmer was to use a source of power other than his or her own, the farm could be bigger and more productive. What might these sources of power be? These are some of the questions we will answer in this chapter.

Let us think of a farmer as a sort of machine! The farmer eats food, which provides him or her with the energy to live and do work (work is the use of energy to produce movement). We can think of the food the farmer eats as a fuel, just the same as a car needs a supply of petrol as fuel to give it energy! The farmer uses up this energy to produce work in the same way as a car uses up petrol. If a farmer is able to do a lot of work in a day we can say he has a lot of power (power is the rate of doing work), just as a powerful car is one with a big engine!

However, no matter how powerful a farmer may be there is a limit to how much work can be carried out in a day. The farmer **can** eat more food to provide more energy to do more work, but there is also a limit to how much a human can eat!

Figure 5.1
These animals (below) are all used to help with agriculture:
(a) camel
(b) donkeys
(c) horse
(d) oxen.
Which are used in Sierra Leone?

Have you ever done a really hard days work in the open air; you probably felt tired and very hungry afterwards. You ate more because you used up a lot of energy!

To increase production, a farmer can use the help of his family or he can hire extra labour; these extra hands can increase the productivity of the farm. However, as humans, their ability to do work is limited.

To be more productive, farmers must look at different forms of energy to provide them with the necessary power to make their farms more profitable.

Animals at work on farms

In early times, people found that large animals were able to do much more work than humans. People tamed (or domesticated) these animals and used them to increase production.

The use of animals remains an important part of African agriculture. The animals that were tamed for this purpose were very suitable for helping humans as they used a cheap and plentiful form of energy, or fuel, in the form of grass. The animals were therefore cheap to run, easy to maintain and even had the advantage that they reproduced themselves. You could not say that about a car or a tractor!

In Figure 5.1 you will see pictures of some of the animals that are used in Africa to help with agriculture. They are usually called 'draught' animals; draught is a word meaning pulling.

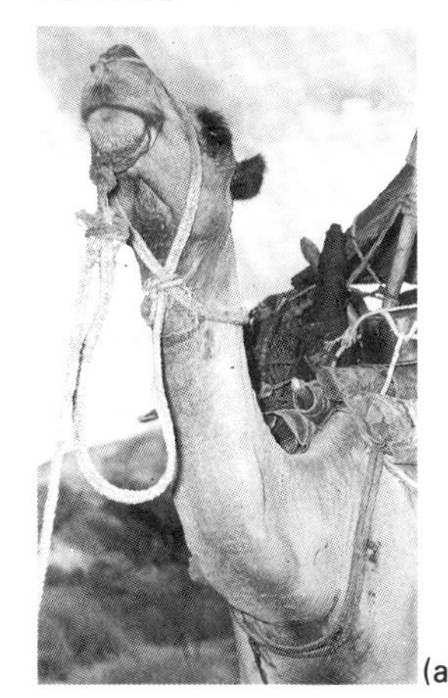
(a)

(b)

(c)

(d)

Which of these animals are used in Sierra Leone? What jobs do they do?

Figure 5.2 Oxen ploughing a swamp for rice cultivation, near Port Loko, Sierra Leone

Oxen

The ox is a most important animal in the agriculture of Sierra Leone; it is with good reason that it has been called the 'living tractor'.

This animal is strong, well built and able to live on poor quality forage (food for cattle, horses, and sheep). If oxen are well fed, properly trained and managed, they become good draught animals (Figure 5.2).

Oxen are very convenient sources of power and are capable of doing a number of jobs on the farm. For example, they can be used for tilling the soil, for pulling light ploughs or carts, for raising water from wells and ditches for irrigation, and for operating small threshing and rice milling machines.

While oxen need to be trained, they can be looked after quite easily by someone who does not have a great deal of experience. Farmers in Sierra Leone should think seriously about the advantages of owning oxen.

If a farmer has two oxen they should be able to work about 7 hectares a year, providing the soil is not very heavy and the land is not too hilly. Perhaps we could think of oxen as being equivalent to men in terms of the work they do. How many men do you think would do the same work as a pair of oxen? Perhaps 10 men equal two oxen? Do you agree with this?

ACTIVITY 1

A visit to a farm that uses oxen

Your teacher will arrange a visit to a local farm on a day when the farmer is ploughing or cultivating with oxen.

Watch the oxen at work. Are they able to do the work of several men? Are they doing a good job? Would it be better if the work had been done by hand? How are the oxen attached to the implement?

When the farmer stops to allow the oxen to rest, ask such questions as:

- How long can the oxen work each day?
- How are the animals cared for?
- How are the animals trained?
- What jobs can they do on the farm?
- How many oxen does the farmer have?
- What equipment does the farmer have for the oxen?
- How much is a trained ox worth?

Measure the area worked by the oxen. This is what to do. Take a stick a metre long and measure the length and width of the patch of land; it is much easier if the area is rectangular. (If the area is an odd shape your teacher will help you with the measurement!) Multiply the length by the width to give the area in square metres. Answer the following questions:

- How long has it taken the oxen to do this area?
- How many days will it take the farmer to finish this task?
- How many people would have been needed to do the same work?

Tractors

While modern tractors are a considerable source of power, they are complex machines. We will discuss tractors in a later book

Figure 5.3
A tractor ploughing a field with a disc plough

but we should, at this stage, consider some of their advantages and disadvantages. While most of our farmers would like to own a tractor, we will see that they are not the answer to all of our agricultural problems!

Tractors are powerful and versatile pieces of equipment; in some cases, they may be able to do the work of several hundred men! If fitted with the proper equipment the tractor is able to do a wide range of jobs; it can plough, cultivate, harvest, irrigate, transport goods and so on (Figure 5.3). Given enough fuel and proper maintenance, the tractor will not get tired, stop for rests or go on strike for more money!

However, tractors have a lot of disadvantages. They cost a great deal to buy, operate and maintain. If not maintained by a skilled mechanic they will break down and spare parts are costly. Tractors use valuable petrol or diesel fuel; at least oxen can eat the grass along the side of the road!

Small horticultural tractors and cultivators

Many designs of these machines are available and, as they are able to do a variety of jobs, they can be a useful aid to farmers.

Some of the best machines are those steered by handlebars so that the operator has good control (Figure 5.4). These machines are supplied with their own sets of implements.

They are expensive to buy and run; perhaps they may be suitable for use in a co-operative (where several farmers may own the machine).

Figure 5.4 Cultivating a swamp with a hand-held cultivator before planting rice

Stationary petrol or diesel engines

Stationary engines are made in many sizes and find many uses on farms; they may be used for generating electricity, threshing

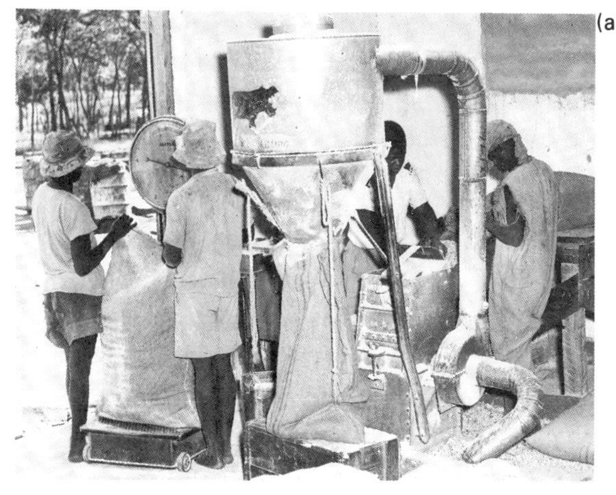

Figure 5.5
Stationary engines:
(a) a village Hippo Mill for grinding maize
(b) a diesel borehole pump

(removing the grains from the ripe heads of crops), grinding corn or pumping water (Figure 5.5).

Other sources of power

Solar energy

Solar energy refers to energy that comes from the sun. The idea of using solar power is very attractive as sunshine is free!

Solar power is very important in agriculture! We have seen that plants use solar power to convert carbon dioxide and water into sugar; do you remember what this process is called? Farmers use the heat from the sun to dry their crops; for example, rice is spread out on the floor to dry it to a low moisture content for safe storage.

However, if we want to trap the sun's energy for a special purpose we would find it quite difficult! The equipment used to

Figure 5.6
These solar panels provide enough power to operate a water pump in a deep well

trap sunshine is very expensive and, unless we have some way of storing this energy, its use is limited to the daytime (Figure 5.6). Why is this? As there is also not much sunshine during the rainy season, the use of solar energy is restricted during this period. Expensive equipment is available that uses solar energy to heat water and generate electricity.

If we wish to save solar energy for use at a later time we must turn it into electricity and store it in batteries; this is even more expensive! Solar-powered lighting, although still expensive, is becoming cheaper.

Figure 5.7
A wind-powered pump being used for irrigation

Wind

When the sun heats up the air it causes air currents; these currents are a potential source of energy. In many countries where winds blow regularly, this energy is used to generate electricity, raise water or grind grain (Figure 5.7).

In Sierra Leone you may have seen a windmill built on a metal frame. The rotor of the windmill is connected by rods to a water pump in a well; when the wind blows, the moving rotor operates a pump and raises water up to the surface.

Wind power is of little value to the small farmer who cannot afford the large cost of the equipment, but its potential value to a village in supplying water is considerable.

Water power

In many countries of the world, moving water, either from rivers, streams or waterfalls, is a source of power (Figure 5.8). (You should remember that the water is moving because, at one time, it fell as rain; the sun caused the water to evaporate, which made the rain. So water power, and indeed wind power, are closely related

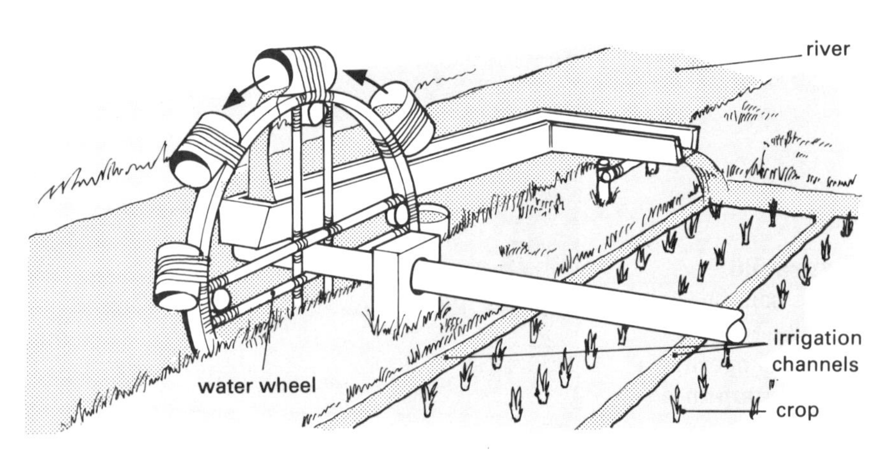

Figure 5.8
A water wheel can be used to lift water from a river into irrigation channels for crop production

to solar power. These types of power are known as sources of 'renewable energy'. Why is this?)

As the water moves or falls from an outcrop of rock it carries weight and force. This energy can be used to generate power. If the water is made to hit against paddles or drop into buckets of a water wheel, the wheels will rotate. The power generated may turn mills for grinding grain, generating electricity or raising water for irrigation. Electricity produced from moving water is known as hydroelectric power.

ACTIVITY 2 Sources of power in your locality

This is a class exercise. With your teacher, make a list of all the different forms of solar, wind and water-powered equipment in your district. Make a note of where they are and what work they are doing. Go and visit each one.

Coal

Millions of years ago forests grew over large areas of the Earth's surface. In time, the remains of these forests were covered with sediments and, due to rock movements and earthquakes, they were buried. Under great weight, the remains were compressed to form the fuel we know as coal. This is dug up and used as a source of energy. This type of energy is known as a fossil fuel and is non-renewable.

Oil and gas

Oil is found deep in the Earth; it was thought to have been formed about 250 million years ago and to have originated from marine life. It is found deep underground and because of its liquid nature it occurs in spaces between porous (loose) rock. It is pumped to the surface where it is refined into the fuels and lubricants (oils and greases) that we use in cars, tractors and lorries.

Useful fuel gases are often found in association with oil. These natural gases and those made when oil is refined are compressed and become liquids. The gases are sold in steel cylinders or 'bottles'. They have to be handled with care and kept cool, for they can explode.

These strong round containers are sold in shops throughout Sierra Leone. They are used for home cooking and lighting. In agriculture, gas is used for lighting and heat in hatcheries and young stock rearing farms. Gas is economic (cost effective) and easy to use.

Oil and gas are also fossil fuels and are non-renewable sources of energy.

Figure 5.9 The methane produced in a biogas digester can be used for cooking, lighting or generating electricity

Biogas

Biogas is a word given to gas which is made from organic matter. It is produced when organic matter is mixed with water and allowed to rot down in the absence of air. The special airtight container in which biogas is produced is called a digester (Figures 5.9 and 5.10). A colourless gas called methane is produced.

The methane is used for cooking, lighting or for generating electricity. A high quality organic fertiliser is produced as residue. In years to come, biogas could play an important part in rural life in Sierra Leone. If a farmer can afford to make or buy a simple digester, the methane can be used for cooking and lighting and the remaining fertiliser can be used on the fields. A biogas producing system is in use at Pa Lokko near Waterloo.

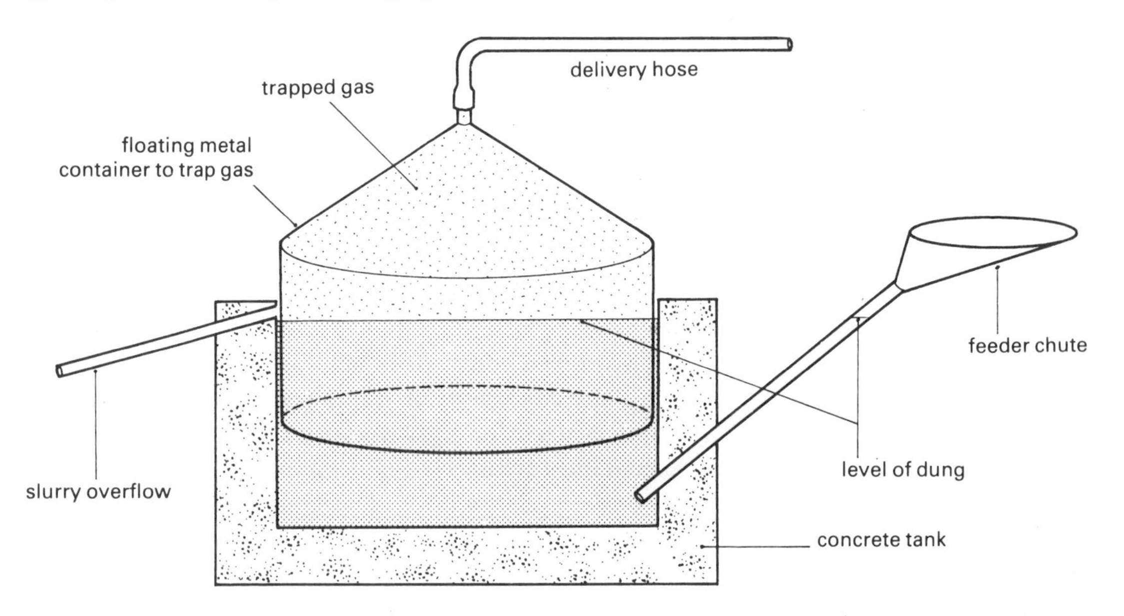

Figure 5.10 Construction of a biogas digester

Sources of power for your farm or garden

We have looked at a range of sources of power and energy that are available to farmers. Which are most suitable for agriculture and which are most likely to be available to our farmers?

Despite the fact that hand work is slow and hard it has some advantages. The farmer will be working close to the ground and will be able to do accurate work, picking out weeds and cultivating between plants. The farmer will not have the worry and expense of buying, running and maintaining machines that may stand idle for much of the year. Much of the hand work carried out in Sierra Leone is done by teams of people who talk and sing together as they work. You cannot do that on a tractor!

Oxen certainly have a place in our agriculture and more farmers should think about owning them. They are useful animals and they are suited to our farming systems. Oxen can help with ploughing and tilling the soil. Their food is grown nearby and it costs little or nothing. While they may be slower than machines, they are quicker than doing things by hand! The farmer still has time to see that the work the oxen do is to his liking.

We have seen that machinery can provide many benefits to agriculture and can, in some cases, replace many men or oxen. The machines can reduce the amount of hard and tedious work in agriculture. However, we have seen the disadvantages of these machines with their high purchase price, high fuel cost and the difficulties in maintenance. Machines have a place on very large farms or in a co-operative.

Some key words and terms in this chapter

Power A measure of the rate of doing work. A kilowatt and a horsepower are measures of power.

Energy The capacity to do work.

Work When energy is applied to something and causes movement.

Draught A term that refers to animals used for work.

Forage A term applied to the grassy foods eaten by cattle, sheep, goats etc.

Co-operative A group of people acting together to all their benefit.

Threshing The process of beating the grains from the ripe heads of plants.

Solar Relating to the sun. As in 'solar energy'.

Renewable energy A source of energy that does not end, e.g. wind, water or solar power.

Non-renewable energy Energy that cannot be replaced once it has been used, e.g. coal and oil.

Fuel Any substance burned as a source of energy, e.g. diesel, petrol or gas.

Lubricant A substance used to make machines run smoothly, e.g. oil or grease.

Biogas A gas produced by the rotting down of organic matter in the absence of air.

Digester The name given to the container in which biogas is produced.

Methane The gas produced in a biogas plant.

Exercises

Missing words

1 The missing words in the sentences below are to be found in the word search. The words may be written horizontally or vertically. Copy the word search into your exercise book and then circle the correct answers.

a ..... power is able to turn a large wheel slowly in a mill.

b The .... from the ... can be concentrated to heat water in pipes.

c The rotor turns when there is a .... and water is pumped up from a well.

d A .... is a powerful piece of agricultural equipment.

e The .. has been called the ...... tractor.

f Working all day on your farm requires much physical ....

B T S T R B X I M
N S T R E N G T H
A L I V I N G R E
G O L Z V W K A A
H A L T E D G C T
E W I N D A C T D
S U N D A Y H O X
T P G W A T E R S
W E L L T R I E D

Multiple choice questions

Write the correct answer to each of the following questions in your exercise book.

2 The main source of power on farms in Sierra Leone is

a wind c engines
b people d water

3 Hydroelectric power is produced by

a wind c water
b solar energy d animals

4 Which of the following animals are used to produce power on the farm?

a pigs c oxen
b sheep d goats

5 Solar energy is obtained from

a water c engines
b sun d wind

6 Biogas is made from

a petrol c soil
b organic matter d methane

Crossword puzzle

7 Copy the grid into your exercise book and see if you can answer the following questions.

Across

1 The gas produced in a biogas digester (7)
2 Work derived from the sun's energy is called this (5,5)
3 A draught animal (2)
4 A measure of the rate of doing work (5)
5 A solid, non-renewable source of energy (4)

Down

6 Power is taken from rivers, streams or waterfalls with this (5,5)
7 It is used to provide the energy for lorries, cars and tractors (4)
8 A big machine used in agriculture (7)
9 When energy is used up to produce movement this is done (4)

Points for discussion

8 Why do you think hand work remains so important in the agriculture of Sierra Leone?

9 What are the advantages of farmers owning a machine co-operatively? What problems would the manager of the co-operative have in keeping all the owners of a machine happy?

CHAPTER 6 The classification and propagation of crops

Introduction

The stage has now been reached for us to look at the plants we use as crops. In this chapter we will find that crops can be described in many different ways. We will also look at the ways in which crops can be propagated (reproduced).

How do we describe our crops?

If you listen to farmers or agriculturalists talking you will hear them use a variety of terms to describe the crops they grow. At first, all these terms may seem confusing but you will quickly come to understand them and to use them yourself. If you are going to talk to farmers, it is important that you understand these terms.

You will discover that crops are described in three general ways:

- in terms of their seasonal growth habit
- in terms of the plant families to which they belong
- in terms that generally describe the crop

A particular crop will, of course, fall into one or more of these categories.

Classification according to seasonal growth habit

In this important classification, farmers and agriculturalists describe their crops according to how long it takes crops to complete their life cycles. Within this group we can recognise three important divisions (annuals, biennials and perennials). We shall look at each one in turn.

Annuals

Annual plants are those plants that survive from one growing season to the next as seeds only. At the start of the growing season,

Figure 6.1 Four important annual crops:
(a) sunflower
(b) maize
(c) tobacco
(d) beans

the seed germinates (starts to grow), the plant develops, flowers are produced, seeds are set and the rest of the plant dies off. The new seeds may then remain dormant (at rest) until the next growing season.

In some cases the whole germination–reproduction process may last only a few weeks and, if growing conditions are good, several generations can be produced each year. This is the case with some weed plants. How do you think this can cause problems for farmers?

Some of the most useful crop plants in the world are annuals. Examples of useful annual crops are rice, wheat, maize, beans, peas, flax, okra, tobacco and sunflowers (Figure 6.1).

Biennials

Biennial plants require two years to complete their life cycle. Biennials do not flower in the first year but develop large food storage organs. Food from these organs is used in the next season to produce flowers and seeds; the plants then die.

Biennial plants are more common in colder parts of the world than in Sierra Leone.

Perennials

Perennial plants are able to persist (continue to live) from one growing season to the next. Perennial plants may be divided into two groups:

- herbaceous perennials (herbaceous means soft and fleshy as opposed to woody)
- woody perennials

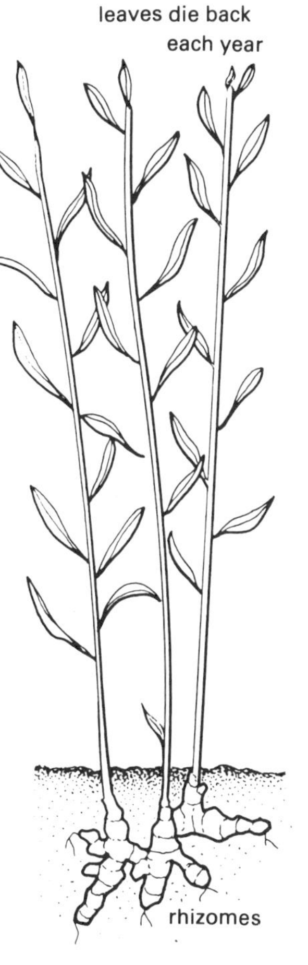

Figure 6.2 Ginger is a herbaceous perennial

Herbaceous perennials

Herbaceous perennials are plants that do not die between growing seasons but persist from year to year; they usually develop some underground storage organs to help them survive. During a period of dormancy (reduced growth due to dry season), herbaceous perennials may either continue to have a few leaves above the ground and grow only very slowly, or their leaves may die down completely and they rely on storage organs in the soil for their survival.

When growing conditions are favourable, herbaceous perennials are able to grow rapidly and produce flowers and set seeds.

Many important agricultural crops are herbaceous perennials; examples are water chestnut, ginger, garlic and some grasses (Figure 6.2).

Woody perennials

Woody perennials are trees, shrubs and vines in which the trunk and branches persist and grow from year to year (Figure 6.3).

In the dry season, some tropical trees may shed some or all of their leaves; these trees are called deciduous plants. The flamboyant is an example of a deciduous tree. However, most trees growing in Sierra Leone do not shed all of their leaves at the same

Figure 6.3 Examples of woody perennial crops:
(a) coffee
(b) pawpaw

time; these are said to be evergreen plants. Bougainvillea is an example of an evergreen shrub.

You would be correct in thinking that many important agricultural crops are woody perennials. Coffee, cocoa, cashew nuts, sisal and mango are a few examples; can you think of twenty others?

ACTIVITY 1 Crop identification

When you next visit a farm, identify each of the crops being grown by the farmer. In your exercise book write down the name of the crop and state whether it is an annual, a biennial or a perennial; in the case of perennials write down if it is a herbaceous or a woody perennial. Your teacher will see if your answers are correct. How may crops did you identify? Were there more annuals than perennials?

Classification according to families

This is not a very important classification for farmers, but one which is often used by scientists.

Plants have been classified into plant families with each family containing plants that are closely related. For example, wheat, rice, millet and maize are cereal crops (plants which produce edible grains) that belong to the plant family Gramineae (Figure 6.4). Pasture grasses also belong to the Gramineae family. Crops belonging to this important group are sometimes called 'gramineous crops' or 'graminaceous crops'.

Another important plant family, as far as agriculture is concerned, is the Leguminosae. A legume refers to the typical dry seed pod produced by all members of this family (Figure 6.5). This family contains plants that are able to manufacture nitrogen

(a)

(b)

(c)

Figure 6.4
Cereal crops such as:
(a) millet
(b) wheat
(c) rice
belong to the plant family Gramineae

compounds using nitrogen taken from air in the soil. This is done in the plants' root nodules by special nitrogen-fixing bacteria. Included in this family are peas, beans, soya beans and groundnuts; scientists will often refer to these plants as 'leguminous crops'.

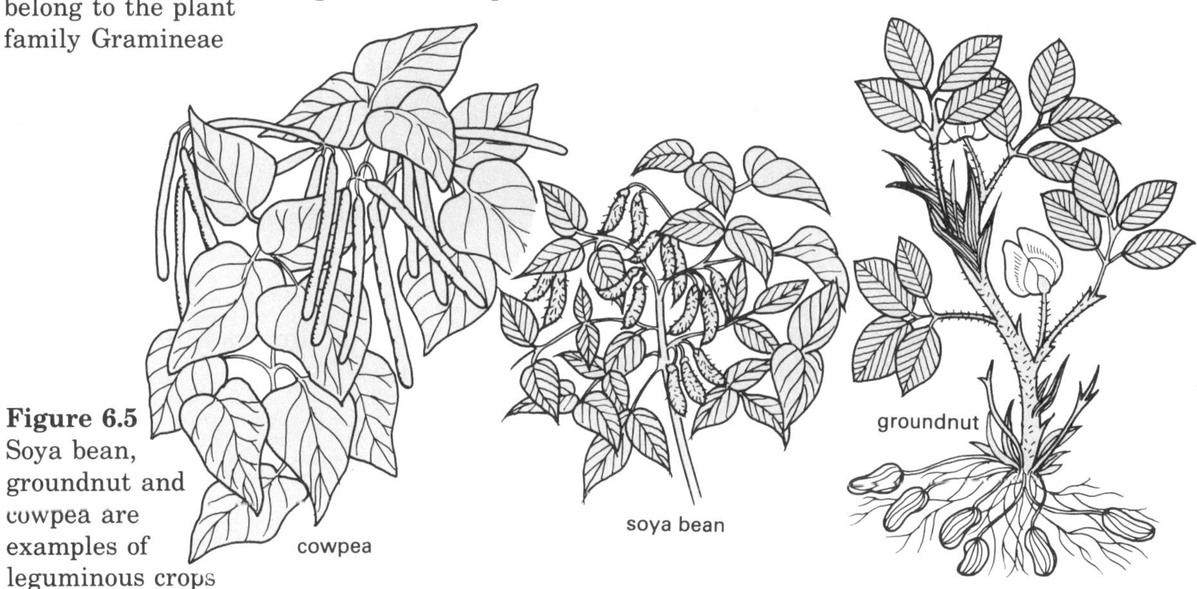

Figure 6.5
Soya bean, groundnut and cowpea are examples of leguminous crops

Descriptive classification of crops

This is an important area because farmers and agriculturalists use a whole range of terms to describe the crops they grow. A list of some of the more important terms follows; examples of each type are given. Can you think of other suitable examples? Remember, it is quite possible that the same crop can belong to one or more groups. It is also possible that a crop which appears in one category may, in fact, have other important uses; for example, we may call cassava a root crop because it has useful roots but in Sierra Leone we grow cassava for its leaves as well as its roots!

Food crop

A crop which is grown to provide food for humans. Examples are peas, maize, millet and pineapple (Figure 6.6).

Figure 6.6
A food crop is any crop grown to provide food for people, such as:
(a) sweet potato
(b) sorghum
(c) pineapple
(d) watermelon

Root crop

A crop grown for its valuable roots. Examples are yams, Irish potato, cassava and carrots (Figure 6.7).

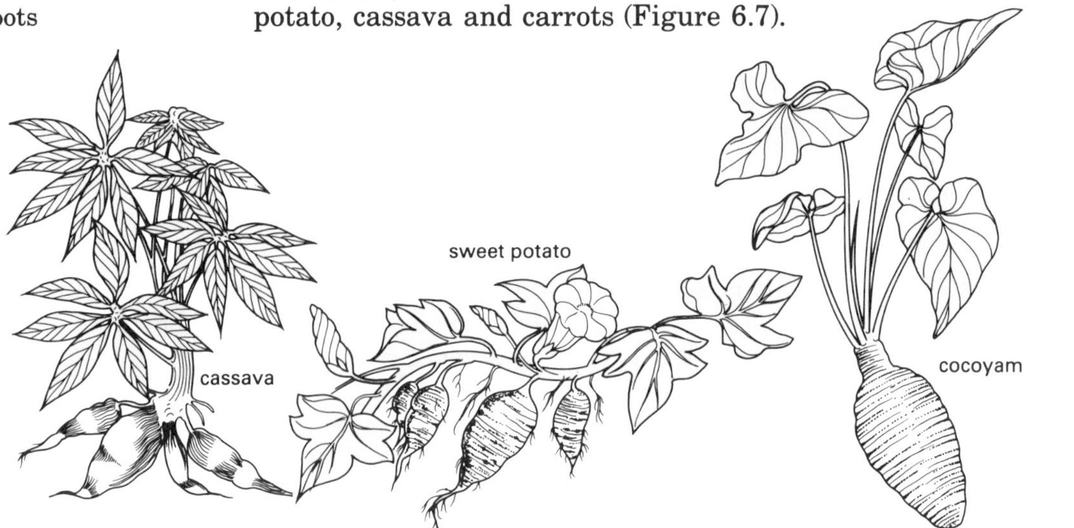

Figure 6.7
Root crops, such as sweet potato, cassava and cocoyam are grown for their valuable roots

(a)

(b)

(c)

Figure 6.8
Examples of vegetable crops:
(a) okra
(b) onion
(c) lettuce

Vegetable crop
A herbaceous crop which is grown for food; these are often grown on a small scale in the kitchen garden. Examples are peas, lettuce, spinach, onion and okra (Figure 6.8).

Plantation crop
A crop grown over a large area. Plantation crops often provide either raw materials for a manufacturing industry or a product for export. Examples are oil palm, coconut, rubber, cocoa, cashew nut and coffee (Figure 6.9).

Figure 6.9
An oil palm plantation

Tree crop
A woody perennial crop which may be grown for fruit, timber or another product. Examples are oil palm, mango, rubber and cocoa. Such crops, which last for many years, may also be called 'permanent crops.'

Forage crop
A crop grown to provide food for cattle, sheep or goats. Brachiaria grass, Guinea grass, *Pueraria* and giant star grass are examples of forage crops.

Cash crop
A crop grown by a farmer with a view to selling it overseas for foreign currencies. Examples are oil palm, cotton, coffee, cocoa and ginger (Figure 6.10). These crops are exported to overseas countries for much-needed foreign currencies. Your teacher will tell you about the role of the SLPMB (see page 15).

Figure 6.10
Cocoa is mainly grown as a cash crop for export

Seed crop
Any crop grown specially to provide seed. The growing of seed crops is a very special business as they must be kept free from weeds, pests and disease.

Shade crop
A crop grown to provide shade for a smaller more delicate crop growing underneath. Tree cassava, banana plants or natural forest trees are used to provide shade for developing crops of cocoa and coffee.

Cover crop
A cover crop serves a similar purpose to a shade crop. A cover crop is sown along with a second more delicate crop to provide it with protection until it is established. This term also refers to a crop, generally a legume, grown as soil cover for erosion control, for example *Pueraria* and *Centrosema.*

Field crop
A crop grown on a large scale as opposed to a crop grown on a small scale in gardens (Figure 6.11). Examples are rice, wheat and pineapples.

Figure 6.11
Pineapples are often grown on a large scale as a field crop (right)

Figure 6.12
Sisal is an important fibre crop (below)

Fibre crops
These are crops grown for their fibres, which may be used for weaving, basket making or various industrial uses. Examples are cotton, sisal, kapok and raffia (Figure 6.12).

Beverage crops
A beverage is a drink. A beverage crop is therefore one in which part of the crop is used to make a drink. Examples are tea, coffee and cocoa.

Crop reproduction

We have just discussed how farmers describe their crops. It is now time for us to think about what farmers use in order to establish (grow) these crops. For us to do this, we must first understand that plants are able to reproduce themselves by two distinct methods. These are vegetative reproduction and reproduction by seeds. We will consider each method in turn.

Vegetative reproduction

Vegetative reproduction occurs when an entirely new plant grows from a vegetative part of a parent plant. The newly formed plant will be complete, with roots, stems, leaves and flowers. We shall find out that a whole range of plant parts may be used in vegetative reproduction.

We should also understand that a plant produced by vegetative propagation is biologically identical to its parent. By reproducing crops vegetatively, farmers are able to grow whole areas of biologically similar plants. What advantages do you think this has for the farmer? Can you think of any possible disadvantages?

Vegetative reproduction is very important in agriculture. Many of our crops are reproduced in this way. Sweet potatoes, sugar cane, pineapples and bananas are examples of crops most commonly grown by vegetative reproduction (Figure 6.13).

Figure 6.13
Cassava is grown from stem cuttings. New plants can be seen growing from the cuttings which have been planted in ridges

Reproduction by seeds

For reproduction by seeds to occur, parent plants must produce flowers. These flowers need to be fertilised with pollen from another flower of the same species. When fertilisation is complete, seeds develop and mature. These seeds can be used to produce a new generation of plants.

The new plants that arise from seeds are not the same as either parent but are a mixture of them both. Just in the same way as you are a mixture of both your parents! Therefore, by growing crops from seed that has been saved each year, a farmer cannot hope to keep a crop quite the same from year to year.

Figure 6.14
Many plants are grown from seed. The picture shows oil palm seedlings growing in a nursery. These are later transplanted into the field

Reproduction by seeds is, however, very important for agriculture (Figure 6.14). Many crops can only be reproduced by seed. Maize, wheat, rice, peas, beans, carrots and okra are some examples. Scientists rely on reproduction by seeds to produce new varieties of crops; it is only by crossing (breeding) one plant with another of the same species that plants with new characteristics can arise.

Propagation of crops in the field

Let us now look at some practical ways in which farmers establish crops in the field and look at the plant materials used to do this. First we will consider vegetative propagation and then we will look at reproduction by seeds.

Vegetative propagation

There are two types of vegetative propagation:

- natural vegetative propagation
- artificial vegetative propagation

Natural vegetative propagation is a form of reproduction that occurs naturally. Farmers exploit (use) this natural process to reproduce their crops. We shall discover that a number of vegetative parts of the plant may be involved in this process (creeping stems or stolons, suckers, underground stems and shoots, tubers, corms, bulbs).

Artificial vegetative propagation includes a number of methods of reproduction in which the farmer must play an active part in the propagation.

Natural vegetative propagation

Creeping stems or stolons

Plants reproducing in this way produce stems or stolons that creep on the surface of the ground; these are also known as runners (Figure 6.15). At intervals these runners produce adventitious (false) roots and buds. The buds produce new shoots and the internodes (length of stem between the buds) may die off leaving several new plants independent of one another. When the internodes do not die off, the result is a series of plants linked by the runners. Carpet grass and sweet potato reproduce in this way.

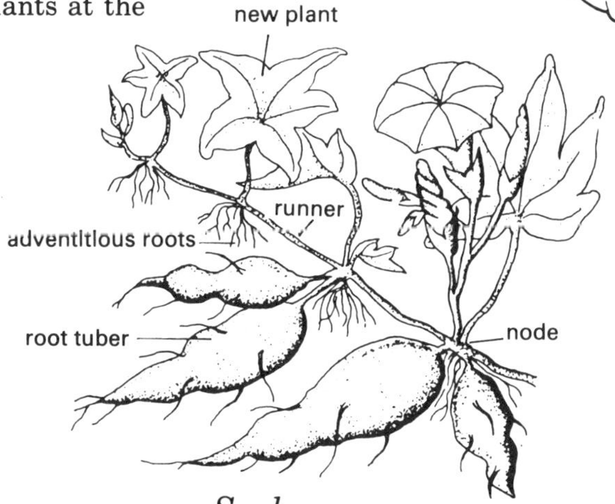

Figure 6.15 Sweet potatoes reproduce vegetatively by producing runners that grow along the surface and form new plants at the nodes

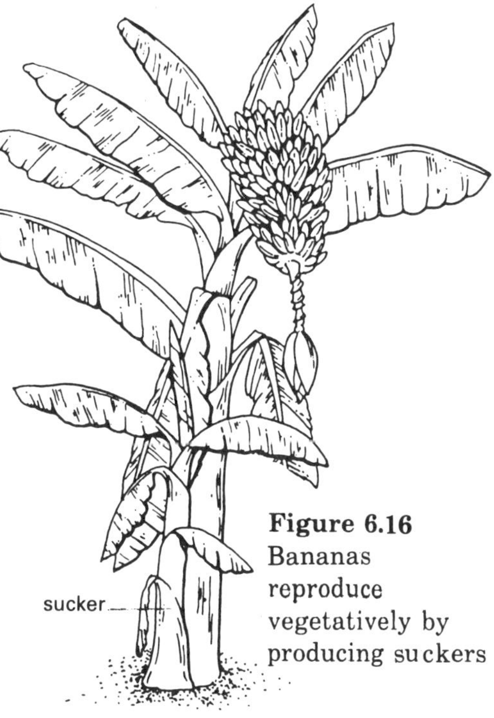

Figure 6.16 Bananas reproduce vegetatively by producing suckers

Suckers

When certain plants mature, young shoots can be seen growing up beside them. These shoots are part of the parent plant as they start from below ground level. When these shoots are old enough to be transplanted they are called suckers. Suckers occur in chrysanthemums, plantains and bananas (Figure 6.16). Farmers reproduce these crops by transplanting the suckers.

Underground stems and shoots

A wide range of crops may be vegetatively reproduced using underground stems and shoots. Most of these organs contain stored food which can be used to propagate new plants. Three types of underground stem are rhizomes, tubers and corms. Bulbs are underground shoots.

Rhizomes

A rhizome is an underground stem which is swollen along its length; rhizomes are generally branched. The rhizome continues to grow from year to year and the bud at the end of the underground stem turns up and grows above the ground to form the leaves and flowers in the current year. Food is passed down from the leaves to be stored in the rhizome. Adventitious (false) roots grow out from the base of the rhizome.

Farmers may propagate new plants from rhizomes by breaking off sections of the rhizome and planting them in a new position.

Ginger and arrowroot are examples of crops propagated from rhizomes (Figure 6.17).

Other plants, notably grasses, produce rhizomes, but these do not act as food stores, but simply as a means of spreading the grass; roots and buds are formed at intervals along the rhizome and shoots grow from the buds, which pierce the soil surface to form a new plant. In this way grasses can spread over an area of land to form dense mats. Brachiaria grass is particularly effective in spreading in this way.

Tubers

Some plants produce buds at the base of the main stem, which then grow laterally (sideways) below the surface of the soil. Food

Figure 6.17
A rhizome of ginger

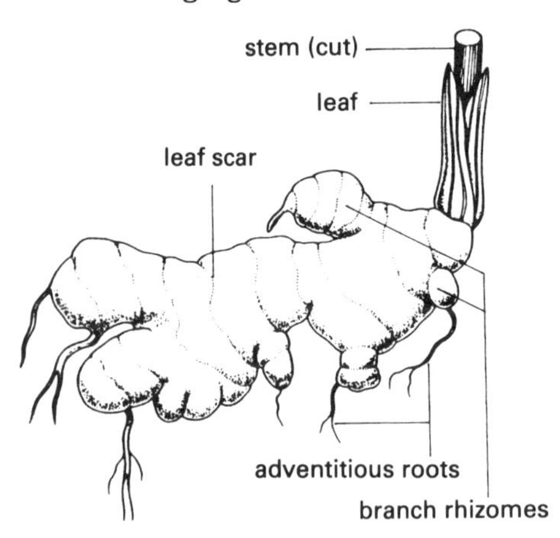

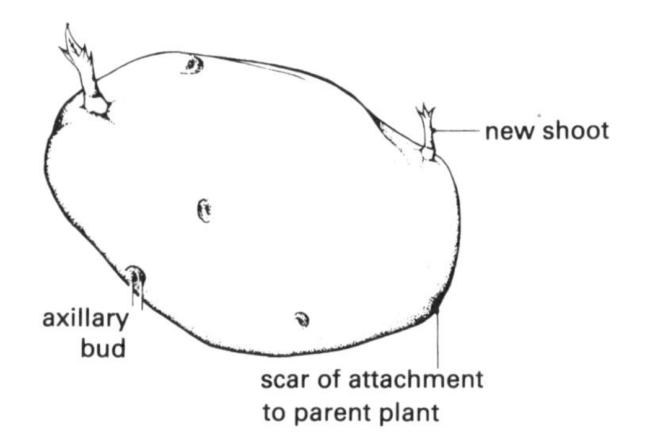

Figure 6.18
Irish potatoes reproduce vegetatively from tubers

from the leaves passes down the main stem to these lateral stems. The ends of these stems swell to become round or oblong food stores called tubers. Tubers are very similar to rhizomes except that the stems do not swell along their entire length.

Each tuber is capable of producing a new plant (Figure 6.18). As the new plant grows the tuber gives up its food reserves and shrivels up. This young plant is now able to produce a new generation of tubers.

By planting tubers from a previous crop a farmer is able to propagate a new crop; Irish potatoes are a very good example of crops propagated in this way. Yams may also be propagated from small tubers; these are often called 'setts'.

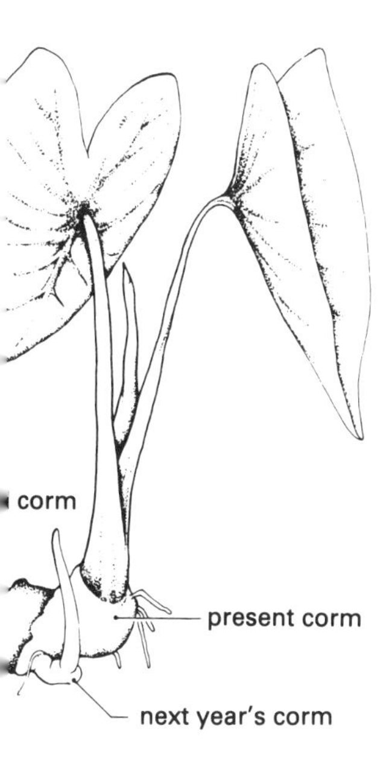

Figure 6.19
Cocoyam is a crop that grows from corms

Corms

A corm is formed when a plant stores food in the swollen base of its stem. Corms are generally round and flattened from top to bottom. The corm is protected by dried outer leaves which form a scaly covering.

Each year young plants and new corms are formed on the side of the parent corm.

The cocoyam is an example of a crop that grows from corms. From Figure 6.19 you will see that the corm of the cocoyam plant is an underground stem with many roots. When this corm sprouts, it grows into a plant exactly like its parent. Other examples of crops that grow from corms are gladioli and water chestnuts.

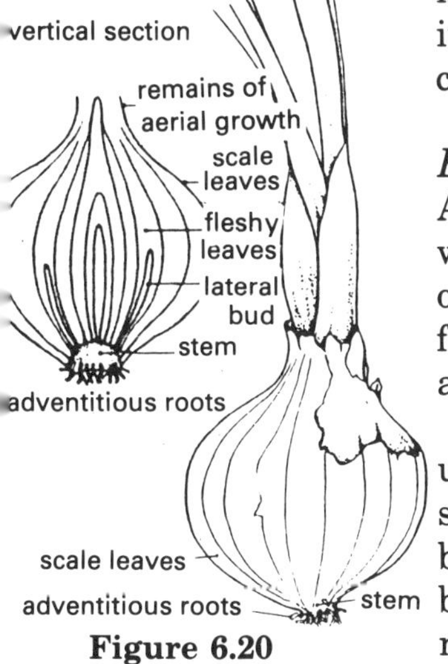

Figure 6.20
An onion bulb

Bulbs

A bulb is a condensed (shortened) shoot with fleshy leaves. You will see from Figure 6.20 that the growing point is in the centre of the structure and is well protected by the outer leaves which form the food storage area. The leaves on the outside have dried, and they protect the bulb.

When a bulb is planted and begins to grow, the food store is used up so that the fleshy storage leaves shrivel. During the new season of growth, some of the food made in the leaves is passed back down the stem to create a new bulb and several new daughter bulbs inside the remains of the old one. Each bulb will create a new independent plant in the next growing season. Onion, garlic and lilies are examples of plants that reproduce in this way.

ACTIVITY 2 Vegetative propagation

Your teacher will arrange for each of you to bring in a particular type of vegetative material. The class will then have available a large selection of stolons, corms, bulbs, rhizomes and tubers.

See if you can identify which is which and what plants they come from; to do this you will have to study them carefully! Your teacher will tell you how many you have got right.

Draw pictures of each of the plant specimens in your exercise book. You may have to cut some of the specimens in half; your teacher will help you with this. With the aid of this book, label your drawings to show all the plant parts.

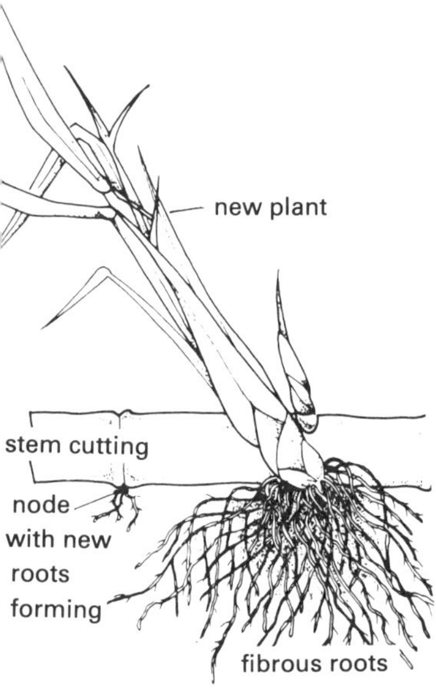

Figure 6.21
A cutting of sugar cane

Figure 6.22
Stem cuttings of sugar cane are placed in trenches and covered with soil

Artificial vegetative propagation

We have learnt that artificial vegetative propagation involves some human activity and that the processes involved do not occur in nature.

Several techniques may be used to create new generations of plants.

Cuttings or slips

Cuttings or slips are terms used for small pieces of parent plants that are removed and placed in the soil and allowed to grow. If conditions are favourable, the cutting or slip will produce roots and develop into a new plant, which is biologically identical to its parent.

Cassava, sugar cane and sweet potato are examples of crops that may be propagated in this way (Figures 6.21 and 6.22).

If a cutting is to 'take' successfully (start to grow) then it is best to include at least one node (a place from which a leaf can grow); it is from nodes and from the cut end of the stem that new roots will develop (Figure 6.23).

Pineapples are a special crop which may be propagated from slips. At the base of the fruit of the pineapple plant you may find two or three buds which have started to sprout; we call these 'slips'. You will see in the diagram (Figure 6.24) that the pineapple has a crown and side shoots which can also be used for propagation. Do not confuse them with slips.

Figure 6.23
Cuttings grown in nurseries are sometimes dipped in a special powder which helps them to form roots (left)

Figure 6.24
A pineapple plant showing vegetative planting material (right)

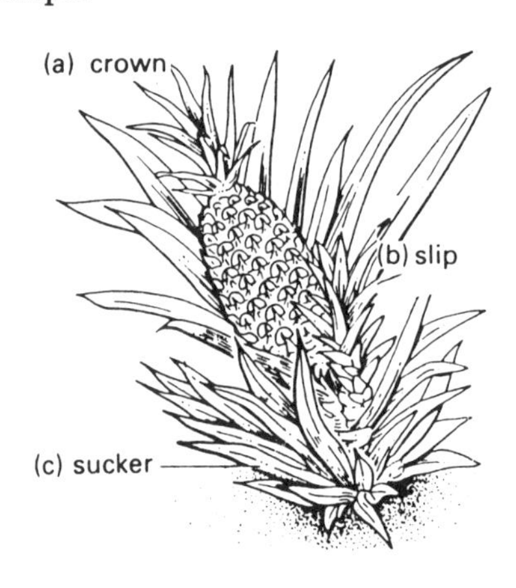

ACTIVITY 3 Setting cuttings

Your teacher will provide you with some lengths of sweet potato stem. Cut off a number of pieces of different lengths; some should include a growing point, others should be middle sections, and some should include one or more nodes. Place the ends of the cuttings in a clear jam jar full of water and allow to stand for a number of days.

After a period of time look at the cuttings and record in your exercise book what you see. How many of the cuttings developed roots? Where did the roots come from? Which cuttings were more successful? Were they the middle cuttings or the cuttings taken from the end of the stem? What conclusions can you draw from this activity?

Layering

This is a method of artificial propagation which may be used for plants that are difficult to grow from cuttings.

The method involves the gentle bending over of a branch of the plant and pegging it into the soil (Figure 6.25). The pegs that hold the branch down may be made from wire or small twigs. Roots should form at the point where the branch has been pegged; a new plant will have been created which can then be cut away from the parent plant.

It is best that at least one node is pegged to the soil; this ensures that roots will develop. Sometimes a cut is made in the branch, close to the node; this also helps roots to form. In layering it is also a good idea to cover the pegged portion of the branch with sand and dead leaves; this will keep things moist until the new plant begins to grow when it can be separated from its parent.

Bougainvillea is a plant that may be propagated by layering; it will not grow from cuttings.

Figure 6.25
The process of layering

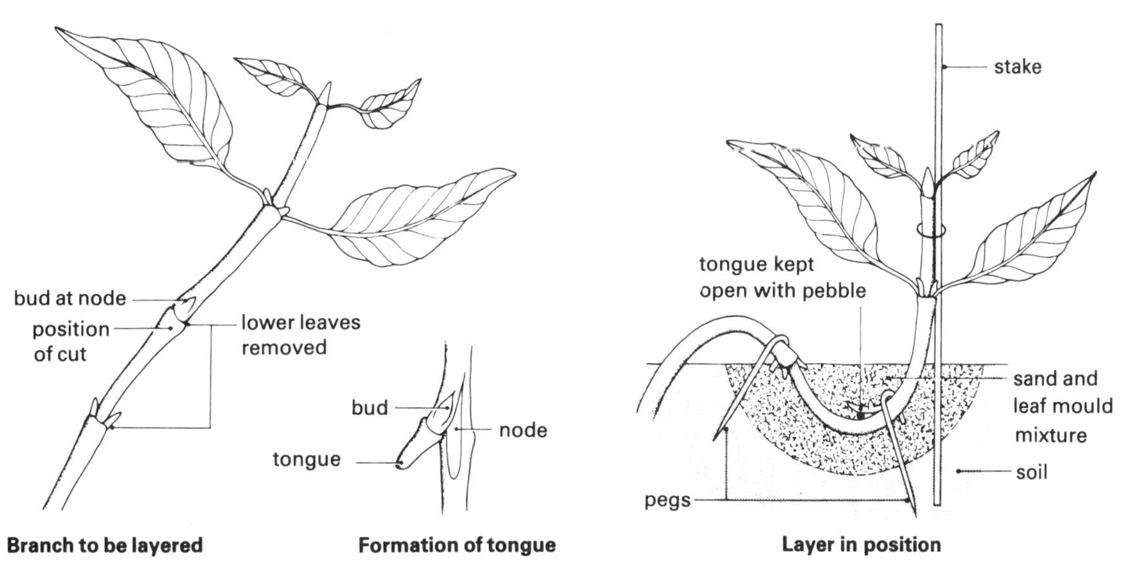

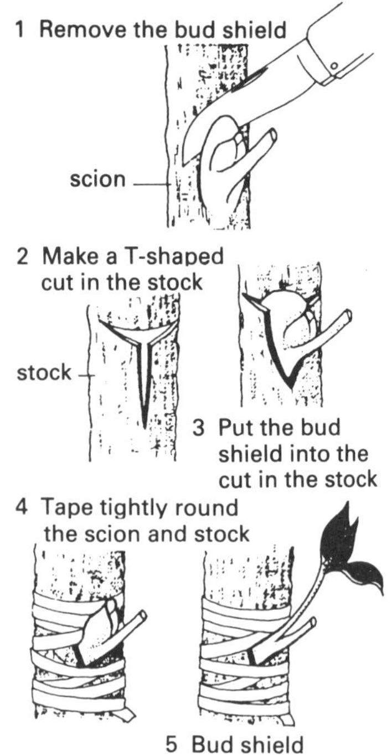

Figure 6.26
Budding

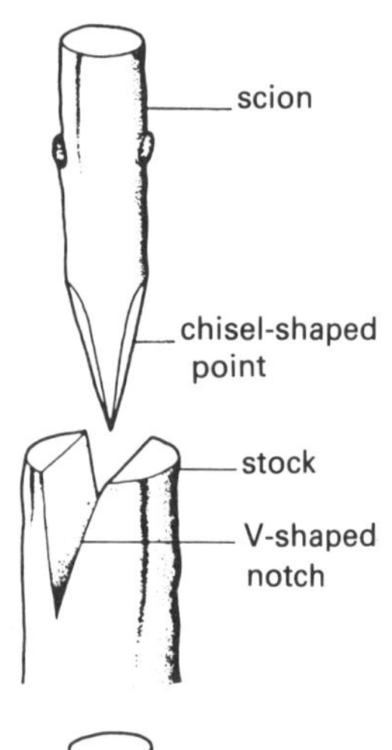

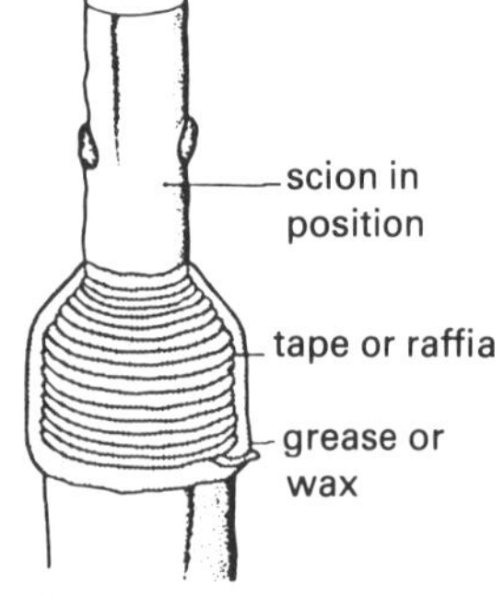

Figure 6.27
Grafting

Budding and grafting

The following techniques of artificial propagation are quite difficult and require a lot of skill. They are methods that are used to propagate woody plants which do not easily grow from cuttings and, in normal circumstances, would not be able to develop strong root systems.

In both these methods a piece of a selected plant is attached to the stem and root system of a strong but closely related plant. The piece of the plant taken to the host plant is known as the 'scion' and the host plant to which it is attached is known as the 'stock'.

Budding and grafting are methods that are mainly used for propagating expensive fruit trees and garden shrubs. Both methods produce plants that are biologically the same as the parent plants. Mango and citrus trees are examples of crops commonly propagated using these methods.

Budding and grafting rely for their success on the fact that the growing areas of the stock and scion are brought into close contact with each other. These methods are best carried out at the height of the growing season when there is the best chance that the growing areas may join together.

Budding is a procedure in which a bud (scion) from a selected shrub or tree is carefully removed and attached to the host plant (stock) (Figure 6.26). The bud should develop and grow into a new plant identical to its parent.

First, a bud from the scion is removed, using a very sharp knife. The bud should be removed on a small square of the bark. A 'T-shaped' slit is then made in the stem of the stock. The selected bud can then be slipped inside this T-shaped cut.

The bud must be kept in place; raffia, string or thin strips of plastic should be carefully wrapped around the newly placed bud. The bud must also be protected from fungal attack, and water loss from the area must be reduced; the bindings should therefore be smeared with a little grease.

Once the scion has taken to the stock and has grown into a strong healthy branch, the remaining branches of the stock may be cut away leaving the new plant to grow on.

Grafting is similar to budding. In this method of propagation a woody twig, rather than a bud, is used as the scion; the twig is specially shaped to fit into a matching cut in the stock (Figure 6.27). While there are several different methods of grafting we will consider only one at this time; we will look at other methods in a later book.

In this form of grafting the base of the twig from the scion is formed into a chisel-shaped point; this is fitted into a matching V-shaped notch in the stock. As in budding, the graft is bound around with tape and protected with grease.

ACTIVITY 4 Grafting

This activity must take place over a period of one year or more!

First, collect together a number of mango seeds; these should be grown in pots or in suitable large tins; we will use these as our stocks.

Once the young trees have grown to 30 cm and when it is the height of the growing season, it is time to practise some grafting.

Select your favourite mango tree to act as the parent plant; remove a number of twigs to act a scions. Prepare the stocks and scions in the manner shown in Figure 6.27, match them up and bind up the graft in the correct way. Make sure you keep the grafted trees well watered. How many of your grafts were successful? Plant your new trees near your home where they will provide you with fruit over many years!

Propagation by seeds

We already understand that farmers use seeds to propagate many of the crops that are commonly grown; in most cases these are annual plants.

For certain crops it is possible for a farmer to save seed to be grown next year; this is commonly done for crops such as rice, millet and maize, where the seed is easy to harvest and the new crops will have many of the same characteristics as the previous ones. In other cases the collection of seed is much more difficult and the resulting crop would be very different. In these cases the farmer may choose to buy new stocks of seed from a seed merchant. The seeds you buy are produced by seed growers; seed production is a very specialised business, which explains why seeds are so expensive. Farmers might, for example, buy new stocks of tomato and carrot seed each year.

Farmers should always sow good quality seed. The following are some characteristics of good seed.

- The seed will be viable (a large number of the seeds will be able to grow). The germination percentage is a measure of seed viability.
- The seeds should be large and firm with a good colour. There should be no broken seeds.
- The seeds will be free from mould and disease.
- The seed sample will not contain any weed seeds or other foreign material.

Until they are ready for use, a farmer should store seeds in a dry place which is free from pests. Seeds in store should be inspected regularly.

WARNING! Many of the seeds you buy from a seed merchant are dressed (coated) with very dangerous chemicals which protect the seeds as they germinate. You should treat such seeds very carefully; always wash your hands after handling them and before eating or smoking; do not breathe the dust. NEVER allow treated seeds to be used for human or animal food.

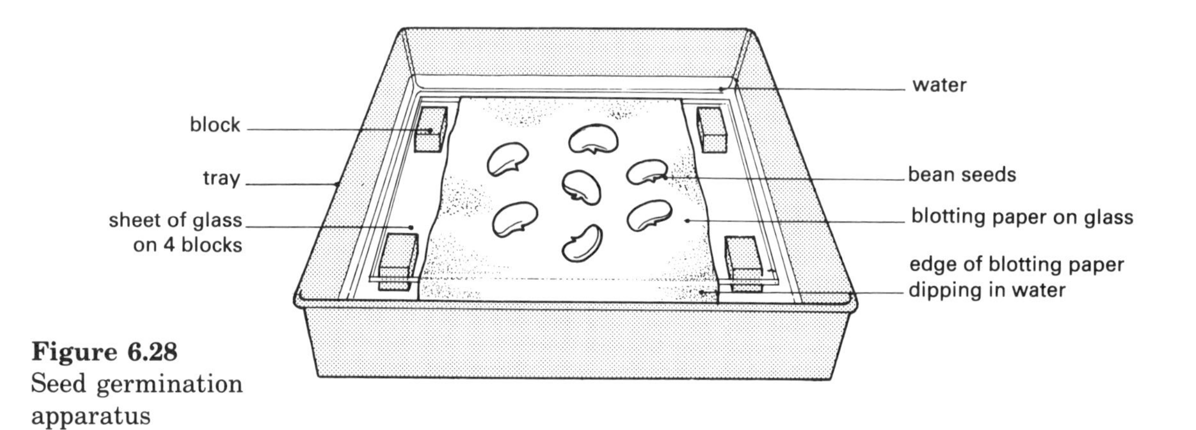

Figure 6.28
Seed germination apparatus

ACTIVITY 5 Seed inspection and testing

For this activity you will need a sample of seeds (rice, maize or beans would be ideal), a tray with high sides, a sheet of glass, some blotting paper or a piece of cloth and some water (Figure 6.28).

1 Inspect a sample of the seeds. What do you see? Does the sample have all the signs of being a good batch of seeds?
2 Count out 100 seeds (you can use 50 or 25 if seeds are limited). Place the seeds on blotting paper which is resting on a sheet of glass, the ends of which are dipping into a tray of water. Keep the blotting paper moist by topping up the water in the tray below. After a period of time the seeds will germinate. Count how many of the 100 seeds have germinated; this is the percentage germination. It should be around 90% to 95% if the seeds were from a good sample!

ACTIVITY 6 Local seed supplies

Make a list of all the places in your locality that sell seeds to farmers; these may be agricultural shops, co-operatives or government stations. Try to visit each one. Which seeds does each one supply? How much do the seeds cost? Where do the seeds come from? Are they produced in Sierra Leone or are they imported?

Some key words and terms in this chapter

Annual A plant which completes its life cycle in one year.
Biennial A plant which requires two years to complete its life cycle.
Perennial A plant which persists from year to year.
Propagate To reproduce or breed.
Vegetative reproduction The propagation of plants without the use of seeds.
Herbaceous This refers to plants which are soft and fleshy as opposed to hard and woody.
Stolon An overground shoot from which new plants arise.
Sucker A shoot which grows up alongside a parent plant.
Tuber A plant storage organ formed from the swollen end of an underground stem.
Sett A term which may be used to describe a tuber, corm or similar plant part used in the propagation of a crop.
Corm A plant storage organ formed from the swollen base of a stem.
Bulb A plant storage organ formed from a swollen shoot.
Cutting A twig or branch which is cut from a parent plant to grow another identical plant.
Slip A term which means the same as a cutting (see above).
Layering Propagating plants by pegging down branches into the soil causing new plants to form.
Budding Propagating plants by taking a bud from one plant and attaching it to the stem and root of another.
Grafting Propagating plants by taking a twig from one plant and attaching it to the stem of another.
Viable A term which is often used when talking about seeds. Viable seeds are capable of germinating.
Germination percentage A measure of the number of seeds in a sample which are viable (see above).
Seed dressing A chemical which is applied to seeds to protect them during storage and germination.

Exercises

Multiple choice questions

Write the correct answer to each question in your exercise book.

1 A stolon is
  a a perennial
  b used for grafting
  c an annual
  d a creeping stem

2 Which of the following terms would **not** be used to describe cotton? Cotton is a
  a plantation crop
  b fibre crop
  c food crop
  d cash crop

3 Which crop produces tubers?
  a coffee
  b banana
  c sweet potato
  d pigeon peas

4 Maize
  a produces suckers
  b is a perennial
  c is used in grafting
  d is an annual

5 Which crop is a perennial?
  a cotton
  b maize
  c cocoa
  d groundnut

Missing words

6 Copy the following passage into your exercise book and complete it by filling in the missing words or terms from the list below.

When you visit a farm you see a variety of crops being grown. Some crops are ...... which means they are sown and harvested in the same year. Other crops may grow on from year to year; these are known as ......

Most annual crops are grown from ...... while many of the persistent crops are ...... by ....

When growing crops, farmers must make sure that the planting material used is of the highest ...... Seeds must be viable and have a good .... All planting material must be free from ...... and disease.

annuals	**germination percentage**
pests	**perennials**
seeds	**vegetative propagation**
quality	**reproduced**

Matching items

7 Match the definition with the most appropriate answer from the list below. Write the answers in your exercise book:

a Soft and fleshy plants.
b A swollen underground stem.
c The pegging down of a branch to the soil to create a new plant.
d Long-living fruit tree.
e The plant that accepts the scion in grafting.

(i) stock	(iv) woody perennial
(ii) layering	(v) tuber
(iii) herbaceous	

Crossword puzzle

8 Copy the diagram into your exercise book and complete the following clues.

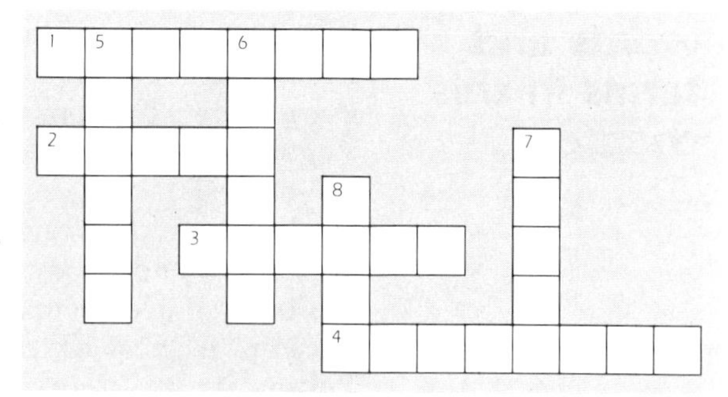

Across

1 Propagating plants by pegging down branches (8)
2 A crop plant which is grown for its bulbs (5)
3 A term used to describe a bean or pea plant (6)
4 A plant, such as carrot, which takes two years to complete its life cycle (8)

Down

5 A plant which completes its life cycle in one year (6)
6 A name for an overground stem which gives rise to new plants (6)
7 The term given to the twig used to produce a new plant in grafting (5)
8 A plant storage organ based on a swollen shoot (4)

Points for discussion

9 Which plantation crops are grown in your area? What are they grown for? Where are the products used? Why are these crops grown in such large areas?
10 What are the advantages and disadvantages of growing large areas of the same crop in one place? This can apply to field crops (such as maize and rice) and to plantation crops (such as oil palm and rubber).
11 Make a list of the important crops grown in your area. With the help of your teacher see if you can put them into their different descriptive categories e.g. cash crops, root crops, fibre crops etc.

CHAPTER 7

Important local crops and plants

Introduction

In the previous chapter we looked at ways in which people describe crops and how farmers propagate them. In this chapter we discover just what a crop is! We will look at weeds and the problems they cause for farmers. We will look at a range of crops commonly grown in Sierra Leone. You will become aware of just how many crops our farmers grow and how important agriculture is in supplying our needs. You will appreciate that farming in Sierra Leone does not just end at the edge of the farmers' fields! Farmers use the surrounding countryside to harvest many things; for this reason we have included some 'crops' which, in fact, grow wild in the bush.

What is a crop?

In previous chapters we have often mentioned crops but we have never really thought about what a crop actually is!

Essentially, a crop refers to a plant or group of plants grown by humans for a special purpose. We have seen that crops may be grown for human or animal food, fibres or industrial products.

You will remember from Chapter 1 that early people collected seeds from wild plants that were found to be useful. These selected seeds would have been planted in specially prepared areas close to the community. By carefully keeping seeds or planting material from the best plants to grow the following year, farmers would have developed varieties of crops that were more useful than the original wild plants. In this way, modern agriculture evolved and plant varieties which suited local conditions were developed. Today, agricultural scientists help to produce even higher yielding crops.

Crop production in Sierra Leone

If we look at crop production in our country we can see that we have two types of farms:

- There are a large number of farms which are managed in a

traditional way; here the farmers cultivate small areas of land, grow a wide variety of crops and do most, if not all, of the work by hand.

- There are a small number of large farms which may be owned by the Government or wealthy individuals. These farms will grow a small number of crops in large areas. Machinery is widely used.

Traditional crop production

In our traditional agriculture, farmers tend to grow a mixture of crops in their fields and gardens (Figure 7.1). There are very good reasons for this! Here are three reasons. Can you think of any others?

- By growing a mixture of crops from different plant families in the same area, a farmer is less likely to have them all killed off by pests or diseases. The organisms that cause such problems are unable to hop easily from one plant to another!
- Our traditional farmers cannot rely on only one or two crops. The farmers must feed their families and cannot risk having serious crop failures; by growing a number of crops, farmers greatly reduce this risk.
- Traditional mixed cropping allows crops to be planted and harvested at different times of the year; this suits the traditional farmer very well as there is only a limited amount of labour available. We saw in Chapter 5 that traditional farmers are not likely to be able to afford much machinery!

We must not be tempted to think of our traditional farmers as being ignorant (lacking in knowledge) or backward in their farming methods. This is not the case at all! Traditional farming systems are very complex (not at all simple) and suit the local needs and conditions very well.

The farmers who work in the fields know a great deal about the work they do and the crops they grow. They are, in fact, **very** knowledgeable and should be respected. Their apparent reluctance to adopt 'modern' farming methods is based on a full understanding of the risks involved!

We must not forget that our traditional farmers rely on many forest trees and bushes to supply their needs. While these plants are not actually cultivated and are not true crops, they are, nevertheless, very important.

ACTIVITY 1 The crops grown by a traditional farmer

Visit a traditional farmer who lives close to you. Walk through the fields where the crops are growing. Make a list of the different

Figure 7.1
A mixed crop of maize and cowpeas (left)

Figure 7.2
Sunflowers growing as a single crop: while such cropping suits modern mechanised farming, crop management must be of a high standard because the crop is open to severe attack by pests and diseases (right)

crops that have been planted and note how the plants are grown; are they all mixed together or are they grown separately? Talk to the farmer about the needs of the household and how these are met during the year. How does the farmer earn some cash to buy the other things that are needed?

In Book 2 we will look in more detail at the layout of a traditional farm.

Modern crop production

On our modern, commercial farms and Government stations you will be able to see large areas of crops being grown; one variety of a plant may be grown over many hectares (Figure 7.2). Growing crops in blocks has many advantages:

- Mechanisation is possible. All the cultivation, sowing, spraying and harvesting can be carried out by machine.
- Large amounts of the crop may be harvested at the same time and can easily be sold or processed.
- Yields can be very high and, given the right conditions, a lot of money can be made.

However, modern agricultural production is not without its problems:

- When we grow a large area of the same crop there is a danger that a pest or disease of the crop can run wild! Farmers may have to rely on chemicals to keep these under control.

- With a large area of crop to be dealt with at the same time, the farmer will need to use machinery. In Chapter 5 we considered some of the problems this can cause!
- It is not easy to grow large areas of the same crop; the farmers need a lot of knowledge and technical help.

ACTIVITY 2 Modern intensive cropping

Visit a large modern farm which is close to you. What crops are grown? Ask the farmer about the problems that have to be faced in growing such large areas of single crops. How are these problems overcome?

Weeds

In the previous section we looked at crops; we found out what we mean by a 'crop' and looked at cropping in Sierra Leone. Let us now think about weeds, which, as we know, are one of the biggest problems when we try to grow crops!

What is a weed?

We can define a weed as 'any plant growing in a place where it is not wanted'! We generally think about weeds as plants growing among our crops which are a nuisance to us and which need to be removed. Weeding is the process of removing these unwanted plants. Have you ever done any weeding?

Why is weeding necessary?

Let us consider why it is important for a farmer to weed crops. Weeding is important because:

- Weeds will compete with the crop for light, air, water and the available food in the soil.
- Weeds often grow faster than crop plants and, unless the weeds are removed, they will smother them.
- Some weeds are parasites; this means that they live off the growing crops. Striga, found in cereal crops, is an example of a parasite.
- Weeds that are removed before they can form seeds will not be able to produce a new generation. However, weed seeds can remain dormant in the soil for many years and it is almost impossible to prevent weeds from appearing in crops.
- The presence of weeds in a crop can cause uneven ripening and difficult harvesting; this is particularly true in intensive modern farming. Weed seeds can be a serious nuisance if they are mixed with the harvested crop.

- Weeding has important secondary benefits. During weeding, a farmer is able to observe the crop closely. Weeding will help to break open the surface of the soil; this will help with drainage and allow air to move into the soil.

ACTIVITY 3

Weed identification

When you visit a local farm ask the farmer and teacher to show you some of the most important weeds that are growing among the crops. Write down the names of the weeds and make a drawing of each of them; this will help you to identify them when you next see them.

The practice of weeding

After planting seeds, it is often very difficult to tell the growing seedlings apart from the weeds which are coming up among them. Unless you are very careful you could loose many valuable seedlings in your efforts to weed! Here are a few hints that may help you.

First look carefully for plants that look exactly the same; these will be the seedlings! The weeds will probably be of a different colour and will have different shaped leaves. If the seedlings were planted in a row it would be helpful if the ends of the row were marked with sticks.

Having identified the crop seedlings it is best to weed by hand around them until they are sufficiently big for a hoe to be used.

Figure 7.3
Weeding cassava by hand and with a hoe

If you use a hoe too close to a delicate seedling you might damage it or disturb the soil and cause it to die.

As you become more experienced you will learn to identify many different crops and seedlings at different stages of their growth (Figure 7.3).

Some crops grown in Sierra Leone

Let us now look at some of the more important crops our farmers grow. We will look at these crops under a series of descriptive headings which we used in the previous chapter. The crops will be described under food crops (to include cereal, legume, root and tree crops), forage crops, fibre crops and crops grown for fuelwood and timber. A short section will then follow where we will look at crops and plants that have medicinal properties.

Food crops

These, you will remember, are crops grown mainly to provide food for humans.

Cereal crops

These are crops that belong to the grass family (Gramineae) and produce edible grains.

Rice

Cultivated rice evolved from the wild rice plant. In Sierra Leone rice is grown in swamp lands or on wet upland soils (see map in Chapter 1, page 18).

After harvest, the paddy is threshed to separate the grain from the chaff. In traditional agriculture this is done by beating the rice with sticks and throwing the separated grains up against a

Figure 7.4
Weeding rice by hand, near Port Loko, Sierra Leone

current of air; the chaff blows away while the grain falls to the ground. In modern farming this whole operation is carried out using a combine harvester.

Further processing must be carried out to remove the husk from the grain. This can be done by pounding the grain in a mortar or by using a milling machine. To make this processing easier, rice is first soaked in water and steamed. The husks swell and open, so that when the grain is dry it may be pounded and the husks fall off easily. The process is called parboiling and it improves the keeping quality of the rice. It also strengthens the grain so that breakage of grain is reduced during milling.

Rice is one of our most valuable staple foods (Figure 7.4). It can also be made into rice beer; the residue is ground into rice meal and fed to pigs and chickens and is an important source of carbohydrate. The husk from rice can be used as litter (floor covering) in poultry houses and as mulch to conserve moisture in soils during the dry season (it is used mostly in pineapple plantations). The husk is also used as fuel for fires.

Sierra Leone used to be the largest producer of rice in West Africa. Much of it was exported.

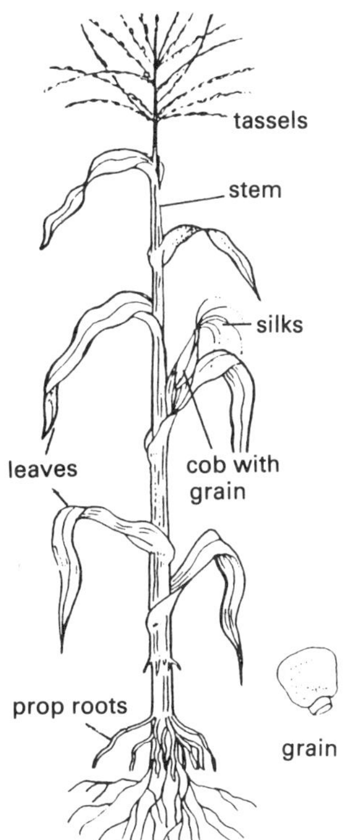

Figure 7.5
Maize plant

Maize

Maize is another important food crop in Sierra Leone. Figure 7.5 shows the structure of the maize plant. You can see the male flowers at the top of the stem with the female flowers, which later form the cob, below. The cob, produced between the leaf and the stem, is a mass of grain surrounding a hard centre.

Maize likes a light sandy loam with plenty of humus. It can be planted on flat beds or on ridges.

The cobs may be roasted and eaten; the grain can be dried and ground to make flour or pressed to provide cooking oil. Maize is a high energy food. Maize can also be used to make maize meal for livestock feeding.

Figure 7.6
A tall white-seeded variety of Guinea corn

Guinea corn (sorghum)
This crop looks quite similar to maize but from Figure 7.6 you will see that the grain forms at the top of the plant. In some varieties the seeds are carried in a compact head (a spike), while in others the head hangs down and is quite open (a panicle).

This crop can be grown like maize, as it likes to grow in a moist soil and it needs a long dry period in which to ripen. It will grow in drier areas than maize.

The grain is harvested, threshed, ground into flour and used in cooking. The flour may also be made into a form of bread or can be used to make beer.

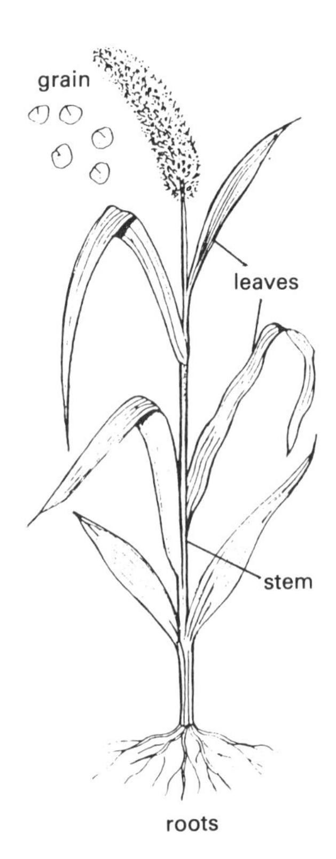

Figure 7.7
Bulrush millet

Bulrush millet
This crop grows mainly in the drier parts of the country. It grows as tall as Guinea corn but usually ripens earlier. The head is thin and tall and the grain small (Figure 7.7).

Millet is made into food for both man and livestock. Like Guinea corn, it can be used in the production of beer.

Hungry rice (fundi or millet)
This crop is not really a rice but is grown instead of rice in places where the land is poor but the rainfall is quite high. It will grow on shallow or rocky soils. If it is sown in June it may be harvested in September. In Sierra Leone it is often grown after rice or as a substitute when the season is unfavourable for rice growing.

Harvesting is similar to the harvesting of rice. The grain is small but has a nice taste. The straw is a good animal feed.

Legumes

We have learnt that plants in this group bear useful seed pods called 'legumes', that they belong to the family Leguminosae and that plants in this family have the ability to give nitrogen back to the soil from their roots. Some of these plants grow close to the soil and form a dense cover while others are climbers and can be grown up sticks. We will look at some of the more important legumes grown by farmers.

Groundnuts
These bushy plants can be identified by their yellow flowers and their large seed pods, which develop at the ends of thin branched stems underneath the soil (Figure 7.8). The legumes (pods) are dug up when ripe; the seeds will store better if kept in their shells.

Groundnuts will grow on almost any well-drained sandy loam but need moderate rainfall and plenty of sunshine.

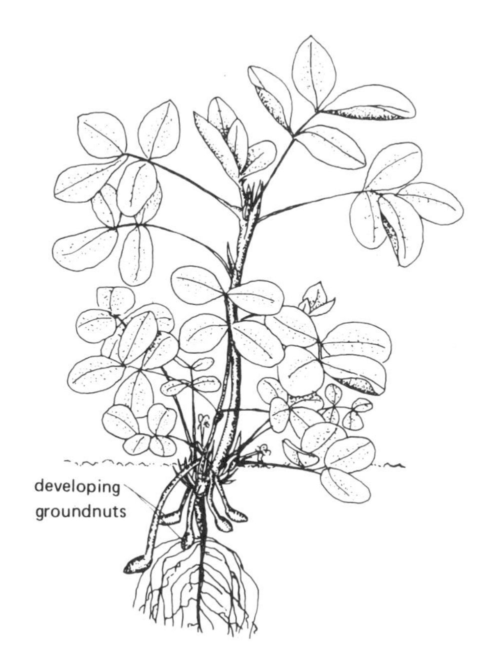

Figure 7.8
A groundnut plant showing the pods developing underground (left)

Figure 7.9
Cowpea plants can be recognised by their long thin seed pods (right)

Groundnuts contain over 50% oil and are a valuable food. After pressing the seeds to remove the oil, the residue is made into groundnut cake. Groundnut cake is a good source of protein and oil for livestock. The seeds can be eaten raw, roasted or boiled. They can also be made into peanut butter which you have most likely enjoyed!

There is a useful export market for groundnuts, but the price depends on the world market prices operating at the time.

Cowpeas

The cowpea is a useful source of food; in Africa it grows wild but is also a cultivated crop. You should be able to identify the cowpea by its long thin seed pods. Study Figure 7.9 carefully and make sure that you can recognise it in both the wild and cultivated state.

Cowpeas grow well in many soil types, even those that are rather infertile. They have a long tap root, which grows down a long way to find moisture; in this way they are able to stand up to dry conditions. Cowpeas are often grown alongside maize or cotton as they replace the nitrogen taken out of the soil by these two crops.

The peas are a valuable protein food for humans. They may be ground into flour for use in cooking. The flour may be made

Figure 7.10
Pigeon pea

into small cakes for sale in the market. Cowpea leaves are a useful feed for livestock; they may be fed fresh to livestock or made into silage along with grass to provide a balanced diet of protein and carbohydrate. Silage is an animal feed which is preserved by fermentation and stored in a silo.

Pigeon peas

The pigeon pea grows in the form of an open bush as you can see in Figure 7.10. The flowers are yellow or orange in colour and produce short, lumpy pods when ripe.

Pigeon peas are often grown with other crops, such as maize or sorghum. The crop grows in the drier areas of Sierra Leone and thrives on sandy loams.

The unripe, green pods are sometimes boiled and eaten. Generally, the ripened peas are boiled and mixed with other food.

Root crops

The plants in this group of crops produce large swollen roots in which food is stored as starch or sugar. Some root crops have to be prepared carefully before eating as they contain poisons; some varieties of cassava, for example, can be very poisonous unless boiled thoroughly. Root crops are usually grown on mounds in soils rich in humus.

Yams

There are many varieties of yam; the one you see in Figure 7.11 has been trained to grow up a support. Study the leaf shape and

Figure 7.11
Yams are usually grown in mounds or ridges, and staked

growth habit of your local varieties of yam as it will help you to identify them.

Unlike cereals and legumes, yams are not grown from seed, but are propagated from small tubers called 'setts'. These setts are planted in the sides of mounds of earth in which the new tubers develop. Harvesting may not take place for 11 or 12 months although some varieties are mature in six months. When the large tubers are dug up they should be stored in a dry house or barn.

Yams can be eaten roasted, boiled, dried and pounded or fried. How does your family prepare yams?

Cassava

The cultivation of cassava is similar to that of yams.The crop is tall and has bare woody stems with all the leaves carried at the top of the plant. Apart from occasional weeding, this crop needs little attention. Harvesting may take place after one year or more; this makes cassava an ideal crop for traditional farmers who may have to rely on it in times of drought or other crop failure. The tubers should be left in the ground until required as they do not store well after being lifted (dug up) (Figure 7.12).

Cassava is grown on a wide variety of soil types and is able to resist dry conditions.

Soaking the tubers in water and then boiling them is the method used to extract the poison that is often found in them. The tubers can be eaten raw, boiled, roasted or fried. When dry they can be pounded to make flour. Cassava is also fermented to make 'foo foo'. Cassava can be used to feed pigs and to produce starch for laundry work. The leaves are an important vegetable in Sierra Leone and are a source of vitamins.

Cocoyam

The cocoyam is a food crop that grows well in wet forest areas. The picture shows you the shape of the leaves and the tall shoot. While the cocoyam is propagated like other yams, the 'setts' that are used are, in fact, corms (Figure 7.13). Cocoyams are usually left in the ground until they are needed. They can also be boiled and made into 'foo foo'. Young leaves and shoots are often cooked and eaten as vegetables.

Sweet potato

The sweet potato originated from South America and was introduced into Sierra Leone many years ago; it has become a very popular crop. You will see from Figure 7.14 that the plant has a creeping habit; the funnel-shaped flowers are usually rose-violet in colour. The leaves, even on the same plant, are not all the same shape so you will have to identify it by its way of growth and its tubers.

Figure 7.12
Cassava plant
(top left)

Figure 7.13
Cocoyam plant
(top right)

Figure 7.14
Sweet potato plant
(below)

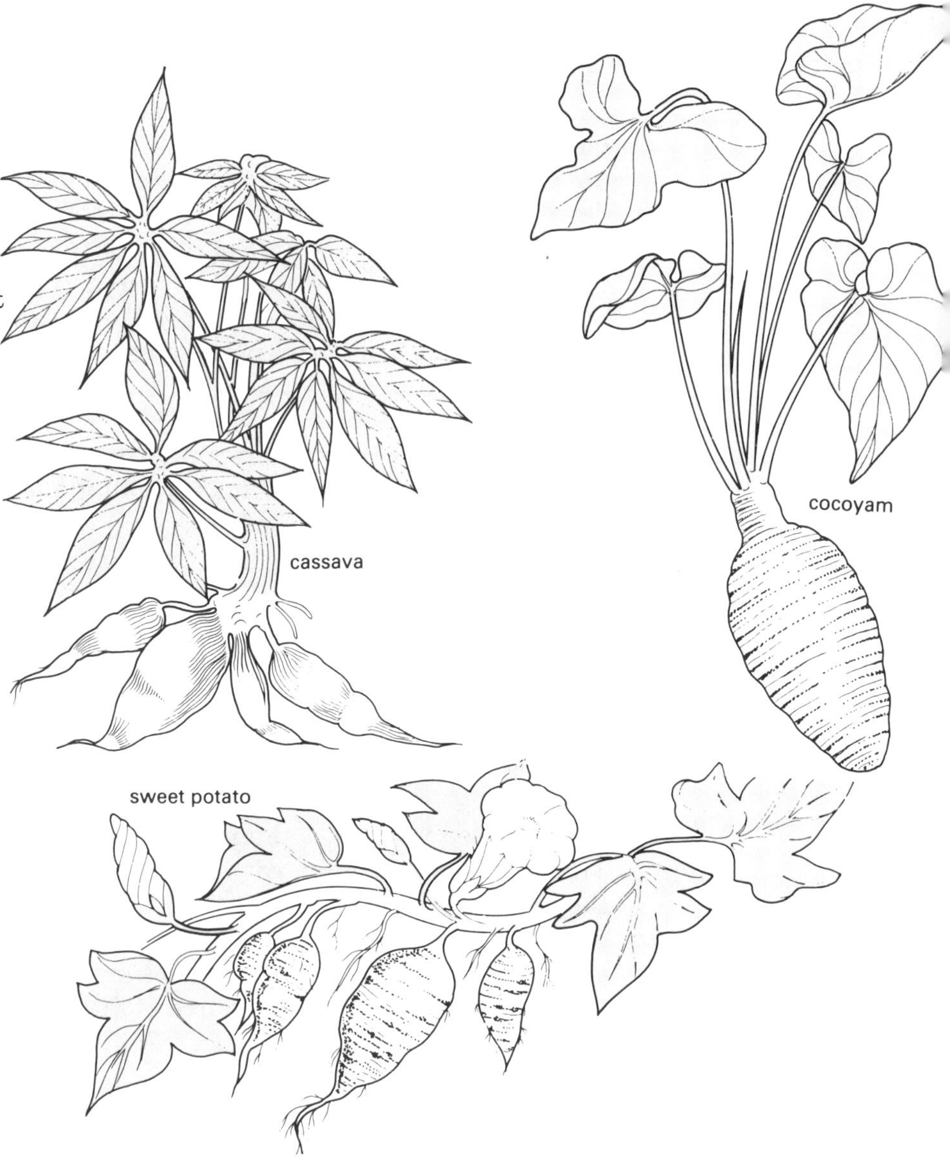

The sweet potato grows in well-drained sandy loams, and likes plenty of sunshine. If the crop is irrigated (artificially supplied with water) it can be grown throughout the year, but it does not like being waterlogged.

The tubers are easily damaged so have to be dried and stored carefully. They are a valuable food as they contain starch, sugar and a small amount of essential proteins. Tubers are usually boiled, roasted, fried or eaten raw. The leaves can also be eaten as a vegetable.

Beverages and stimulants

In this section we will look at a number of plants that provide us with beverages (drinks) and stimulants (something that increases body activity).

Figure 7.15
An orange tree with fruit

Citrus
In the citrus group we include lime, lemon, orange, tangerine, sour orange, mandarin and grapefruit (Figure 7.15). We will use the lime as our main example as it is found throughout Sierra Leone.

The lime is a small evergreen tree or bush producing many small and greenish-yellow fruits. You will probably know the fruits well and enjoy the taste of the juice! If a lime tree is given good growing conditions it will produce high quality fruit. The leaves have a characteristic scent when crushed. Compare the shape of the leaves of a lime tree with those of the orange and lemon; see what small differences there are.

Like other citrus fruit, lime juice has a high vitamin C content, which is valuable in a drink.

Oranges are usually grown in small groves near villages and the trees are tended well. They produce sweet fruit for eating, providing a useful source of energy and vitamins. The sour orange is used for cooking and export. The lemon is less common and, like the mandarin, is of little value due to its sharp taste. Grapefruit are usually grown in small plantations near the main towns and have become more popular.

Palms
Many varieties of palm can be found growing throughout Sierra Leone. They provide us with a range of useful products.

Palms provide us with sap which is obtained by 'tapping' (drawing off from) the top of the tree (Figure 7.16). The sweet sap, called palm wine, is drunk fresh. Care must be taken not to kill the tree by taking too much sap from it.

Coconut palms provide us with a number of useful products. From the nut we obtain a refreshing drink and the flesh of the nut provides us with a useful food. Coconut oil can be extracted

Figure 7.16
A palm tree being tapped for sap (left)

Figure 7.17
Coconuts are harvested by climbing the trunk of the palm and cutting the ripe fruit (right)

from the nuts and the residue makes an excellent animal feed. The husk of the fruit is made into coir which can be used for making mats and rope or can be burnt as a fuel (Figure 7.17).

From the raffia palm we obtain the raffia used in weaving. Palms also provide us with roofing and building materials.

Cola

The cola is a tree that is valued for its nuts. Cola either grows wild in the forests or is cultivated in plantations. Cola plantations are often found in wet forest areas which provide the shade necessary for the young trees. The trees do not produce nuts until they are at least nine years old. The cola grown in Sierra Leone is said to be of very high quality.

The fruit contains a stimulant which reduces feelings of hunger when it is chewed. The nuts are exported to be made into the familiar 'Coca cola' or other similar drinks. Extracts from cola nuts are also used as a dye in gara making.

Coffee

Coffee bushes may be found growing in shady areas with good

rainfall; they do not grow well in strong sunlight. Coffee is grown for the fruits, which are called 'berries'; these must by individually picked by hand (Figure 7.18). The berries are processed, dried and roasted to produce the coffee beans which provide us with a stimulating drink.

Coffee bushes need a lot of attention. Regular pruning can help to improve yields but the bushes will eventually grow too tall and will have to be replaced.

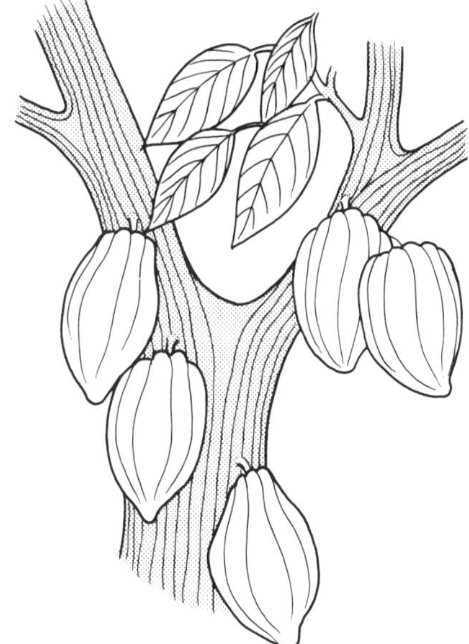

Figure 7.18
A branch of a coffee bush showing berries (left)

Figure 7.19
Cocoa pods growing on the trunk and branches of a cocoa tree (right)

Cocoa

Although cocoa is a native of America, it is now grown in many tropical parts of the world. Countries of West Africa, the East Indies and the Far East are major producers of cocoa. West Africa is the leading producer supplying some 65% of the world market. Cocoa grows in tropical rain forests where conditions are warm and humid. Cocoa is a small evergreen tree about 6 m tall (Figure 7.19).

Cocoa seeds are used for making cocoa powder and chocolate. Cocoa butter is an oil extracted from the seeds which may be used in the manufacture of chocolate, toffee, margarine, cosmetics and pharmaceutical goods. After processing, the remaining portion of the cocoa seeds may be finely ground and mixed with cocoa or chocolate to produce cheaper and poorer quality products. The residue may also be used as a fertiliser or cattle feed.

Forage crops

These are crops that are grown to be eaten by cattle, sheep or goats.

In Sierra Leone we can identify two different ways of keeping livestock. While we will look at these in detail in Chapter 9 it is important, if we are going to look at forage crops, that we have a basic understanding of them now.

First, we have the traditional ways of keeping livestock. Here, the animals are usually kept in relatively small numbers and are allowed to forage for their food in the countryside or around villages. The animals may be free to wander about, they may be looked after by a herdsman or they may be tethered (tied) and allowed to graze a small area. The exception to this rule is to be found in the north of the country where the Fulani people keep large numbers of cattle. These cattle herds are carefully tended by herdsmen who move them around the region following the seasonal growth of feed (Figure 7.20).

Figure 7.20
Fulani herdsman with his cattle

When animals roam around nibbling at plants and trees we often say that they are 'browsing'. The herdsmen in charge of the animals will know what plants the animals like and will take them to where there are a lot of these special browse plants.

Secondly, we have the modern, intensive ways of keeping livestock; these methods are used on large commercial farms and on Government stations. Here, large numbers of animals are kept in a relatively small area. Beef and dairy cattle, sheep and goats will be allowed to graze in fields in which special grasses have been grown or they may be kept in pens where the feed is carried to them (Figure 7.21).

At times of the year when forage is plentiful, the excess may be cut and preserved; this may be fed at other times of the year when there is a shortage.

Figure 7.21
In intensive dairy farming, the cattle are kept in pens where the feed is carried to them

Traditional farming and forage production

It will be seen that our traditional farmers do not really grow crops specially for their livestock. The animals are allowed to eat the plants that grow in the bush and around the fields or villages or they are permitted to eat the residue of crops that have already been harvested for human use; rice left in the field after harvest and sorghum or millet stalks are often eaten by livestock. Traditional farmers usually spend their time growing food for their own use and have little time to grow special forage for their animals. However, certain crops are grown which man and animals may share; crops such as beans and pigeon peas are good fodder for cattle, sheep and goats and may also be used to supply food for the farmer.

During the dry season some traditional herdsmen and farmers set fire to the bush. This destroys the dead material and encourages some plants to send up fresh shoots which can be grazed. While this may be of benefit in that it provides some new grazing and the ash may act as a fertiliser, the soil lies bare and unprotected. The wind may then cause erosion and the topsoil can be carried away. In general, the gains obtained from burning the bush are offset by the possible loss of soil fertility and the killing of useful organisms. Figure 3.11 (page 59) shows 'slash and burn' agriculture.

ACTIVITY 1 Visit to a local livestock herd

Your teacher will arrange for you to visit a local herdsman or farmer who keeps some livestock. You must find out as much as possible about the way in which the animals are fed. What plants do the animals eat throughout the year? How does the herdsman provide the plants? How does the herdsman grow or manage the forage? What do the animals do in the dry season? Who owns the land on which they graze? Does the herdsman have to pay to feed his animals? How far do the animals move during the year? Write down the answers to these and other questions in your exercise book.

Intensive farming and forage production

In intensive farming, a large number of animals must be supplied with a steady supply of forage. This requires a lot of knowledge and skill; the farmer or herdsmen cannot allow the animals to go hungry. Considerable planning must take place to make sure there will be enough forage throughout the dry season.

Special crops will be grown to supply the animals' needs. During the growing season some of the crops may be grazed by the animals in the field. Others will be allowed to grow, and will then be cut and conserved (saved) for use during the dry season.

Figure 7.22 The rear view of a tractor-operated baler making hay (right)

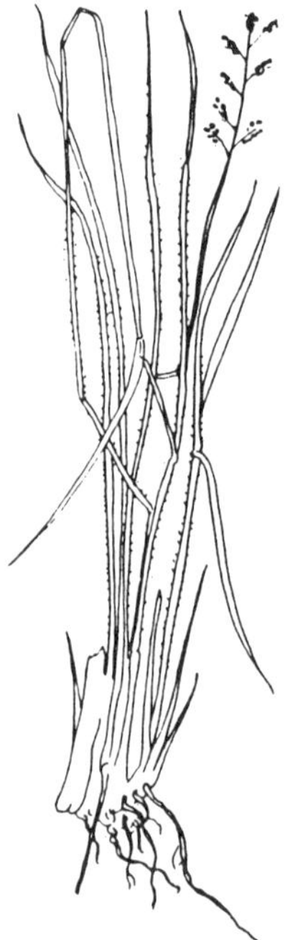

Figure 7.23 Brachiaria grass

Crops may be conserved by drying; dried grass is known as hay. Other crops may be put in a sealed pit and allowed to ferment (preserve themselves); silage is the term used to describe crops stored in this way. The making of hay and silage usually needs a lot of machinery (Figure 7.22).

Cultivation of a typical forage crop

Brachiaria grass is an example of a forage crop grown in Sierra Leone and is particularly suitable for this purpose (Figure 7.23). The grass can be established by transplanting, it is very prolific (it spreads easily over the surface of the soil), it stands up to grazing and trampling by animals and is able to persist during the dry season.

ACTIVITY 2

Forage production and intensive livestock

Your teacher will arrange for you to visit a local farm which keeps intensive cattle, sheep or goats. In this exercise you will find out as much as possible about how the animals are fed and how this system differs from the traditional method of feeding livestock. Ask such questions as: How many animals are kept? What area of ground do they occupy? What forage crops are grown? What are the advantages and disadvantages of each forage crop? How are these crops established (planted and grown)? How much fertiliser is used? How much is used for grazing and how much is used for conservation (making into hay or silage)? How are the crops conserved for the dry season? How much is conserved? How are the animals fed? How much do they get each day?

When you get back to the classroom, discuss with your teacher the differences between traditional and intensive livestock keeping.

Crops grown for fibre

Some plants are grown by the farmer because they contain fibres that are strong enough to make into rope, cloth and other similar items. We can identify two types of fibre:

- *Soft fibres.* These are made into soft materials, such as cloth.
- *Hard fibres.* These fibres are tough and are used for making ropes and matting.

We will examine a few examples of fibre crops that you will find growing on farms and in the bush.

Cotton

Cotton is a crop which is either grown on a large scale to provide fibres for the manufacturing industry or is grown on a small scale at village level where it is used to make country cloth.

The valuable soft fibres of the cotton crop are found in the bolls, which form when the flowers die and the plants set seed. The bolls do not all ripen at the same time, so several pickings have to be made. Figure 7.24 shows the bolls, which hold the fluffy white fibres or lint, being harvested. The bolls contain seeds which are black and hard; the seeds provide us with a useful oil and the

Figure 7.24
Harvesting cotton in Sierra Leone (left)
Figure 7.25
The kapok or silk cotton tree is grown for the fibres that surround its seeds (right)

residue (cotton seed cake) is used for feeding livestock. Before the fibres can be cleaned, spun and woven, the seeds must be removed.

Cotton is mainly grown in the north of the country where there is a long dry season, which is necessary for the crop to ripen.

Raffia palm

This palm, which grows in Sierra Leone and has the largest leaves in nature, provides us with fibres as well as other useful products. While the stems are used for building, making ladders and some furniture, the leaves make good thatch (roof covering). Raffia fibre is excellent if used as a string in horticulture. Piassava or bass is produced by allowing the soft material in the leaves to rot away in running water; the strong leaf veins, like rods, which remain are used in constructing brooms and brushes.

Silk cotton or kapok

Although the silk cotton tree is not strictly grown as a crop it does provide useful fibres and will be included here. Often found growing close to villages, this huge tree produces black seeds which are surrounded by a mass of silky floss or fibres; this is kapok. The tree also has many medicinal properties (Figure 7.25).

Kapok is used for stuffing pillows and cushions because the fibres do not easily flatten. Kapok has also been used for filling the lifebelts carried on ships as the fibres are able to stay afloat for many days. The seeds contain an oil which is used for making soap and for cooking.

Crops grown for fuel and timber

Sierra Leone is fortunate in having a great variety of trees that produce some of the best timber in the world. However, as the population increases, the demand for firewood and building timber will rise and there is a danger that the forests will not be able to replace the trees cut down. This is why it is so important to conserve what is left and to replace trees as quickly as possible. This problem is made worse because, as farms increase in size to meet the demand for food, more and more trees are cleared away. While a few farmers may plant trees to supply fuelwood, few farmers are prepared to plant trees for timber.

In later books we will look at the techniques of agroforestry. Here farmers grow trees alongside their crops. Agroforestry will become increasingly important in Sierra Leone in the future.

Mangium

Mangium is a quick growing, evergreen, leguminous tree, which is often grown by farmers to supply fuelwood (Figure 7.26).

Figure 7.26
Mangium trees are planted by farmers to supply fuelwood

Like all legumes, the mangium improves the fertility of the soil by releasing nitrates, which may be used by other crops. For this reason, it is often grown in association with other permanent crops, such as cashew and coffee.

Acacia
There are a number of trees and shrubs in this genus (group) that provide timber and fuel. The trees are also often browsed by livestock. Acacia trees have an umbrella shape and the feathery leaves give shade over a wide area. When a farm is cleared, the farmer will usually leave acacia trees for this reason. Acacias prefer the semi-arid parts of the country.

Mango
Having eaten its fruit, you will probably know this tree which is usually found close to most villages and towns! It is one of only a few trees that are cultivated by farmers for fruit, shade and wood (Figure 7.27).

Figure 7.27
A fruiting branch of a mango tree

Given the right conditions and a fertile soil, the large seeds grow quickly and the young seedlings establish themselves easily. After a year, the seedlings are transplanted. After a further year, a selected variety should be budded on to the young stock. Mango trees are sometimes planted in rows along streets in towns where they provide shade and where anyone has access to the fruit. The wild mango produces a good timber, which is used in house building, and the seeds contain a useful oil.

Neem
This tree was originally brought to Sierra Leone from India and is well-suited to our climate; it may be found by houses and

Figure 7.28 Neem is a fast-growing fuelwood tree

villages. It grows quickly, which makes it a most useful fuelwood. The leaves may be used for making paints and varnishes, soap and medicines (Figure 7.28).

Mahogany

A number of the hardwood trees in this group grow in Sierra Leone. One variety is found in the drier forest areas and is used in the production of canoes and furniture, and to make planks for house construction. Mahogany is a valuable timber for export.

You may find the seeds of the mahogany on the ground under a tree. As they will germinate easily, push a few into the soil in a moist place nearby. The mahogany is a useful tree for reafforestation.

ACTIVITY 3 Tree recognition and use

Copy the table below into your notebook. Your teacher will take you to a forest area or to a place where trees grow together. Hopefully you will be able to meet a forester or a farmer who knows a lot about trees. As you walk, ask the forester or farmer to give you the names of trees that you see and to tell you of their use. Record this information in your table. Look carefully at the different trees; note their height, colour and type of bark. Look at the leaves; what shape and colour are they? How high is the tree? Record this information in your table as well. On your way back to school, see if your friends have remembered the names of the trees you saw earlier.

Table 7.1 Trees and their uses (Activity 3)

Name of tree	Description of tree	Uses of timber	Properties of timber

Crops with medicinal properties

Wild and cultivated plants were used to cure or treat diseases long before modern medicines were developed; lots of country people still rely on such plants when they are ill. Many of the medicines used by today's doctors come from products discovered in plants. In this section we will look at a few plants with medicinal properties.

Our forests and wild places are home to hundreds of different types of plants; many of these may still be unknown to modern science. The medicinal properties of most of our native plants have yet to be investigated. It is important that we should not destroy our wild plants and that we should leave them for future generations. We must protect our forests and bush as they may hold plants which contain cures to some of our most serious diseases.

A word of warning to you! Do not eat any fruit or plant material that is unknown to you unless your parents say that it is safe. As you know, many plants are poisonous. So beware, and do not experiment!

White yam

The scraped pulp of the white yam tuber is useful when applied to burns as the juice appears to have a numbing effect on human skin.

Commelina

You may know commelina as a weed which causes problems in groundnut cultivation where the pieces, broken by hoeing, tend to regrow into new plants. The leaves of commelina are used in cooking as a pot herb and the stems are used to probe wounds (Figure 7.29).

Sweet potato

The leaves of the sweet potato have an unusual use! They can be used to kill insects. The leaves may be mixed with charcoal and applied to house walls. Insects, as you know, can transmit many diseases, so it is a good idea if they can be controlled around the home.

Figure 7.29 Commelina and prickly amaranth are used as medicinal herbs

Prickly amaranth

The prickly amaranth has value as a pot herb and as fodder for cattle. In some areas it has been used as an ingredient for enemas to treat internal troubles (Figure 7.29).

Sekou touri

The sekou touri is a hardwood tree whose leaves and roots are used to alleviate the symptoms of malaria. It is a tree that grows wild all over Sierra Leone, both in the rainforest and savanna areas. It is also used for fuel.

Pawpaw

Pawpaw trees are grown for their ripe fruits which are eaten. The whole plant contains a white sap, which is capable of tenderising tough raw meat. The bruised leaves wrapped around meat have the same effect.

ACTIVITY 4

Medicinal plants used in your community

Copy the table shown below into your exercise book. When you go home this evening ask your parents, relatives and the old people in the village what plants they use to treat sickness and ailments. When you return to school your teacher will collect together all the information gathered by your class. You will see just how important many of our plants are in keeping us healthy! Who in the village knows most about traditional medicines?

Table 7.2 Medicinal plants

Plant	Description	Where found	Medicinal use

Some key words and terms in this chapter

Crop A plant or plants cultivated by humans for a special purpose.
Weed Any plant growing where it is not wanted.
Parasite An animal or plant that lives in or on another, from which it obtains nourishment.
Threshing The process of separating the ripened seeds from the husks and straw of plants.
Parboil To boil until partly cooked.
Litter The absorbent material used on the floors of intensive livestock houses. A litter is also used to describe a family of young pigs or rabbits.
Spike A seed head which is shaped into a sharp point.

Panicle An open seed head in which the seeds are carried on a branched stem.

Silage A forage crop, which has been stored by fermentation, to be fed to livestock during the dry season.

Tapping The process of drawing off sap from a tree, as in rubber and palm trees.

Browsing The nibbling of trees, shrubs and grass by animals as they move about.

Conservation A term applied to the storing of forage crops as hay or silage for use in the dry season. Conservation, of course, can also mean the caring for our wildlife and environment.

Boll The fruit of the cotton plant which contains the fibres and the seeds.

Agroforestry The growing of trees in association with crops.

Medicinal Having healing properties.

Exercises

Multiple choice questions

Write the correct answers in your exercise book.

1 Which definition best describes a weed? A weed is a plant which
   - a has no value to the farmer
   - b has no bad effects on the farmer's crops
   - c produces lots of persistent seeds
   - d is growing where it is not wanted

2 Which crop returns nitrogen to the soil through its root system?
   - a cocoa
   - b millet
   - c rice
   - d groundnut

3 Which of the following crops is **not** a cereal crop?
   - a pigeon pea
   - b millet
   - c rice
   - d maize

4 Which crop is also known as Guinea corn?
   - a sorghum
   - b maize
   - c millet
   - d rice

5 Which root crop favours wet forest areas?
   - a cocoyam
   - b cassava
   - c sweet potato
   - d white yam

6 The crop that is grown for its berries, which are picked, processed and made into a drink is
   - a cocoa
   - b coffee
   - c cola
   - d citrus

7 What is the alternative name for a fibre producing plant which grows wild and produces kapok?
   - a raffia
   - b silk cotton
   - c acacia
   - d neem

8 What is the name of a medicinal crop whose leaves are used as a pot herb and the stems are used to probe wounds?
   - a sekou touri
   - b prickly amaranth
   - c commelina
   - d white yam

Missing words

9 Copy the following passage into your exercise books and fill in the gaps using the words from below.

Our traditional farmers must, first of all, ... themselves and their families. They are not wealthy and cannot ...... having serious crop ...... For this reason they grow a number of crops. The farmers will usually mix their crops together; this reduces the ...... of ...... and disease. By growing a number of crops the farmers are able to spread the ...... as the crops are ...... at different times of the year. Only when the farmer has enough food for the family can the ...... be sold for ......

surplus	**pests**
workload	**risk**
feed	**harvested**
failure	**cash**
spread	

Crossword puzzle

10 Copy the diagram into your exercise book and complete the following clues.

Across

1 Another name for Guinea corn (7)
2 Describes crops grown for their natural filaments used for weaving etc. (5)
3 Obtained from palms and used for making brooms and brushes (8)
4 A leguminous tree which provides shade, timber and browse for animals (6)
5 A citrus fruit which is good to eat (6)

Down

1 A root crop grown for its tubers. The leaves on the same plant may vary in shape and colour (5,6)
6 Produced from a palm and used for making bags and horticultural string (6)
7 'African ....' is another name given to the answer to 3 across (4)
8 Root crops usually grown on a framework of sticks (4)

Points for discussion

11 We have learnt that our forest trees are being cut down faster than they can be replaced. Why is this a problem? What can be done about it? What happens when too many trees are cut down?

12 Which are more important to the development of Sierra Leone, our traditional farmers or our large commercial farms and plantations?

13 Should our traditional farmers be encouraged to adopt more modern farming methods? How could this be done? What do they need most?

CHAPTER

8

The growing of vegetables

Introduction

An important stage has now been reached as you will be able to grow your own vegetables. In this chapter we will explore the conditions required for seeds to germinate, we will discover how vegetable crops are established and how they must be cared for.

We will grow a crop of local beans as an example of a crop grown from seed. If the beans are to give you a good harvest you will have to look after them with great care. In this chapter you will receive step-by-step instructions on how to look after them. See who is able to grow the best crop of beans in your class!

What is a vegetable?

A vegetable may be defined as 'any of the various herbaceous plants with parts that are used for food'. How many vegetables can you name? There are really quite a number of them!

Let us think about vegetables in terms of the way we use them. Let us see if we can sort them out into different groups. First, we can think of vegetables that we eat because of their roots (carrots, sweet potato). Then, there are those we eat for their shoots (onion). There are those grown for their leaves (spinach, cabbage), and finally, others are grown for their flowers and fruits (cauliflower, eggplant) (Figure 8.1).

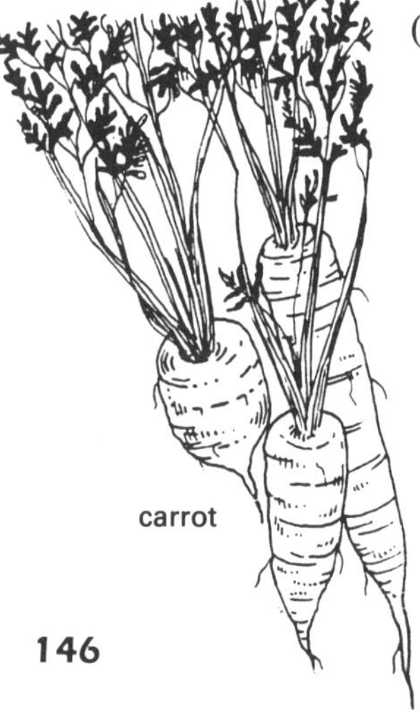

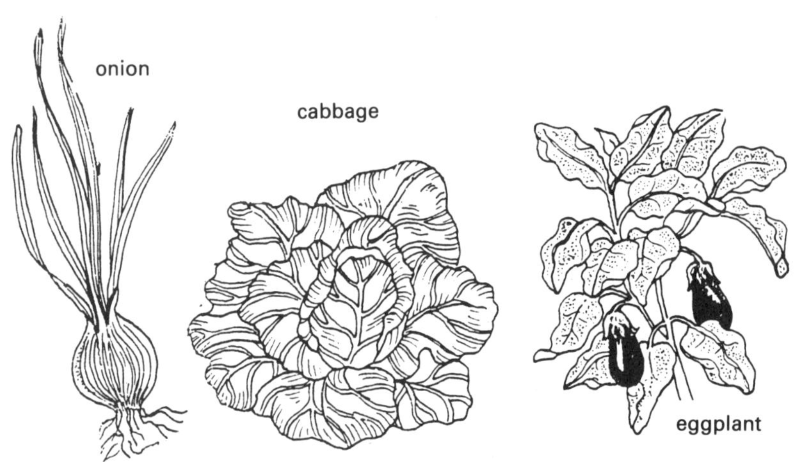

Figure 8.1
Vegetables are grown for their roots (carrots), shoots (onion), leaves (cabbage), and fruits (eggplant)

ACTIVITY 1

Classification of vegetables

Copy the four-part table shown below into your exercise book; leave plenty of room as each section may become quite large!

Below is a list of vegetables. Write their names down under the appropriate heading in the table and describe the part of the crop that we use. A few examples have been completed for you, to help you to start.

Now think of as many other vegetables as you can and put them under their appropriate headings. Your teacher will gather information from all your class to make up an even bigger list! You will see just how many vegetables there are!

Table 8.1 Vegetables and their uses (Activity 1)

Group	Vegetable	Part(s) eaten
Root	Carrot	Orange tap roots
Stem	Onion	Swollen shoots
Leaf	Spinach	Upright green leaves
	Cabbage	Leaves of compact apical bud
Flowers and fruit	Cauliflower	Young florets
	Eggplant	Immature fruits

Examples of vegetables

pepper	radish	thyme	parsley
carrot	water melon	lettuce	peas
mint	beans	tomato	krain-krain
sorrel	spinach	water leaf	okra

Establishment of vegetables

We have learnt in Chapter 6 that our crops may be established in a number of ways and from different types of material. We have seen that crops may be established by vegetative propagation or may be grown from seeds. In this chapter we will look at the growing of vegetables from seeds.

Growing vegetables from seeds

The techniques used in the planting of seeds require our special attention.

We know already, of course, that crops should only be grown from top-quality seeds, which should be obtained from a reliable source. Your vegetable seeds should preferably be bought from a

Figure 8.2
Agricultural seed merchant's shop in Freetown, Sierra Leone

reputable agricultural merchant or garden supplier (Figure 8.2). In Sierra Leone seeds may be bought from such organisations as the Seed Multiplication Project, Home and Garden or the Horticultural Division at New England in Freetown. Check that your seeds have not passed their 'sell by' date.

You should read the instructions on the packet and follow them as closely as possible.

ACTIVITY 2

Inspection of bean seeds

Your teacher will give you about 40 bean seeds which you will plant; these will be of a common local variety. The beans may be harvested to be eaten green, or may be grown for the mature seeds.

Examine your bean seeds closely. Do they seem like a good sample? Are the seeds the same size, colour and shape? Are the seeds 'full' or do they look shrivelled and wrinkled? Are there any cracked or broken seeds in the sample?

Conditions required for germination

Before we plant our seeds it is a good idea to know what conditions the seeds require if they are to germinate (start to grow). Only by giving your seeds the best conditions can you expect to get a good crop.

Seeds require warmth, air and moisture if they are to germinate successfully (Figure 8.3).

- *Warmth.* In Sierra Leone the climate is always warm enough to allow seeds to germinate. However, we must be careful not to allow our seedlings to get too hot as they may wilt and die. It is often necessary to shade delicate seedlings from the sun;

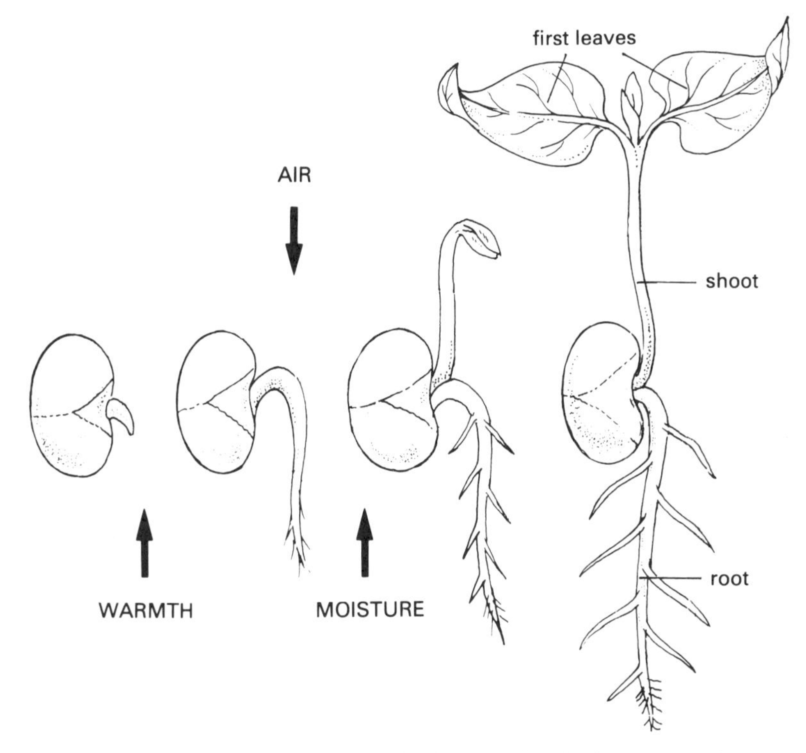

Figure 8.3
Germinating seeds need warmth, air and moisture

this is something we will look at a little later in this chapter.

- *Air*. Seeds are living things and as they grow they need to respire or 'breathe'. We must make sure that our soil is loose enough to allow for air spaces and that it is not waterlogged.
- *Moisture*. As seedlings develop they require a constant supply of moisture. We must make sure that our young plants always have enough water; in the hot sun, seedlings can quickly die if they run short of moisture. We must not, however, overwater our seeds; too much dampness can create the conditions which allow fungi to attack the young plants. When young seedlings die from a fungus attack it is a condition known as 'damping off'.

Seeds generally do not need light to grow. Seeds contain their own reserves of food. Only when the seedlings have grown above the ground do they need light. At this stage they develop chlorophyll, turn green and start using sunlight to produce food.

ACTIVITY 3 The germination of seeds

For this activity you will need four bean seeds, an empty jam jar, some blotting paper (pieces of paper towel, cloth or a napkin would do just as well) and water (Figure 8.4). This is what to do:

1. Draw a picture of one of the beans in your exercise book; your teacher will help you to label it.
2. Roll up the blotting paper and fit it inside the jar; allow it to unroll so it touches the sides.

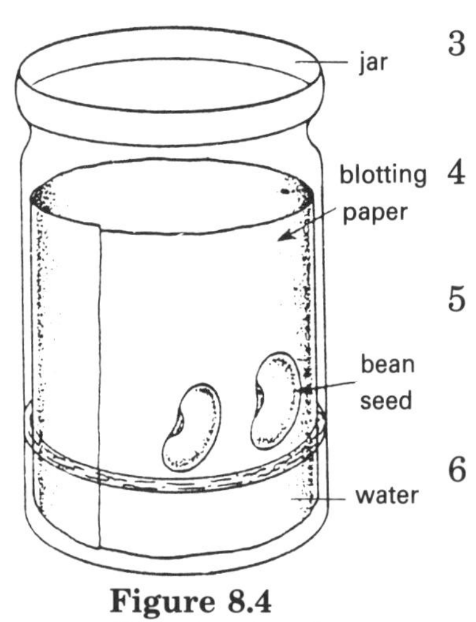

Figure 8.4
Germinating bean seeds

3 Push the four beans carefully down the inside of the jar between the paper and the glass sides, until they are half way down and evenly spaced around the jar.
4 Carefully add water to the jar so that the blotting paper is thoroughly wet and there is a little water sitting in the bottom of the jar.
5 Place the jar in a place which is out of direct sunlight. Add a little water to the jar as required; the blotting paper should be kept moist and there should always be a little water in the bottom of the jar.
6 After periods of 4, 8, 12 and 16 days, carefully look at the beans and see what has happened. Draw pictures of the germinating beans in your exercise book. Your teacher will help you to label your drawings. Discuss what you have seen with others in your class.

Planting vegetable seeds

You will find that vegetable seeds may be sown in a number of ways; the proper way will depend on the type of vegetable. The different methods are as follows:

- Some seeds must first be planted in seed boxes or in a nursery bed and, when they have grown to a certain size, they must be transplanted.
- Some seeds are planted directly out into the garden. Once the seeds germinate there are those which:
  - (a) need not be thinned out; the beans you will grow belong to this group; the seeds may be sown in rows;
  - (b) are usually grown in rows and must be thinned out to the required spacing (excess plants are removed leaving the stronger ones to grow on to maturity);
  - (c) are broadcast and need not be thinned out, for example krain-krain and sorrel.

ACTIVITY 4

Seed description and method of establishment

In this exercise we will look at all the vegetables grown from seed that we identified in Activity 1. We will describe the seeds and we will decide how the crops are established.

Copy down Table 8.2 into your exercise book. Your teacher will help you to complete it. Under 'method of establishment' you are to decide if the crops are:

- planted in a nursery or in seed boxes for later transplanting
- planted out (in rows or broadcast) and need no thinning
- planted out in rows and need thinning

A few examples are given to help you to start.

Table 8.2 Seed planting methods (Activity 4)

Crop	Description of seed	Method of establishment
Beans	Large brown seeds	Sown directly in field in rows; no thinning
Tomato	Tiny brown seeds	Sown in a nursery bed or seed box, then transplanted out in the garden
Carrot	Small curved brown seeds	Sown directly in rows, thinned

You should be able to identify the seeds of many crops just by looking at them. Perhaps your teacher can arrange a quiz in which you are shown different kinds of vegetable seeds.

Growing seeds in a nursery

Small and delicate seeds often need to be raised in a nursery before being transplanted. Only when such plants are big enough will they be able to withstand the harsh environment of the field or garden. Such seeds are either sown in specially prepared nursery beds or in seed boxes (Figures 8.5 and 8.6).

In a seed nursery, the farmer can keep a very close eye on the growing seeds and can give them special attention.

Seed boxes

Rather than grow delicate seeds directly in the ground, many farmers prefer to raise seedlings in seed boxes. While this requires more work, the seeds can be managed more easily and the seeds are given a better start. Seed boxes can be carried from the nursery to the field when it is time to plant out the seedlings.

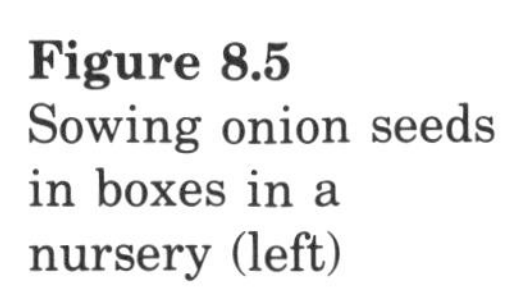

Figure 8.5 Sowing onion seeds in boxes in a nursery (left)

Figure 8.6 A nursery bed shaded with palm fronds (right)

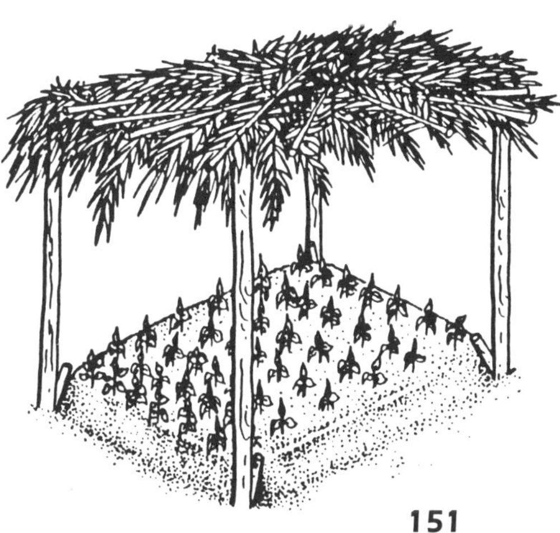

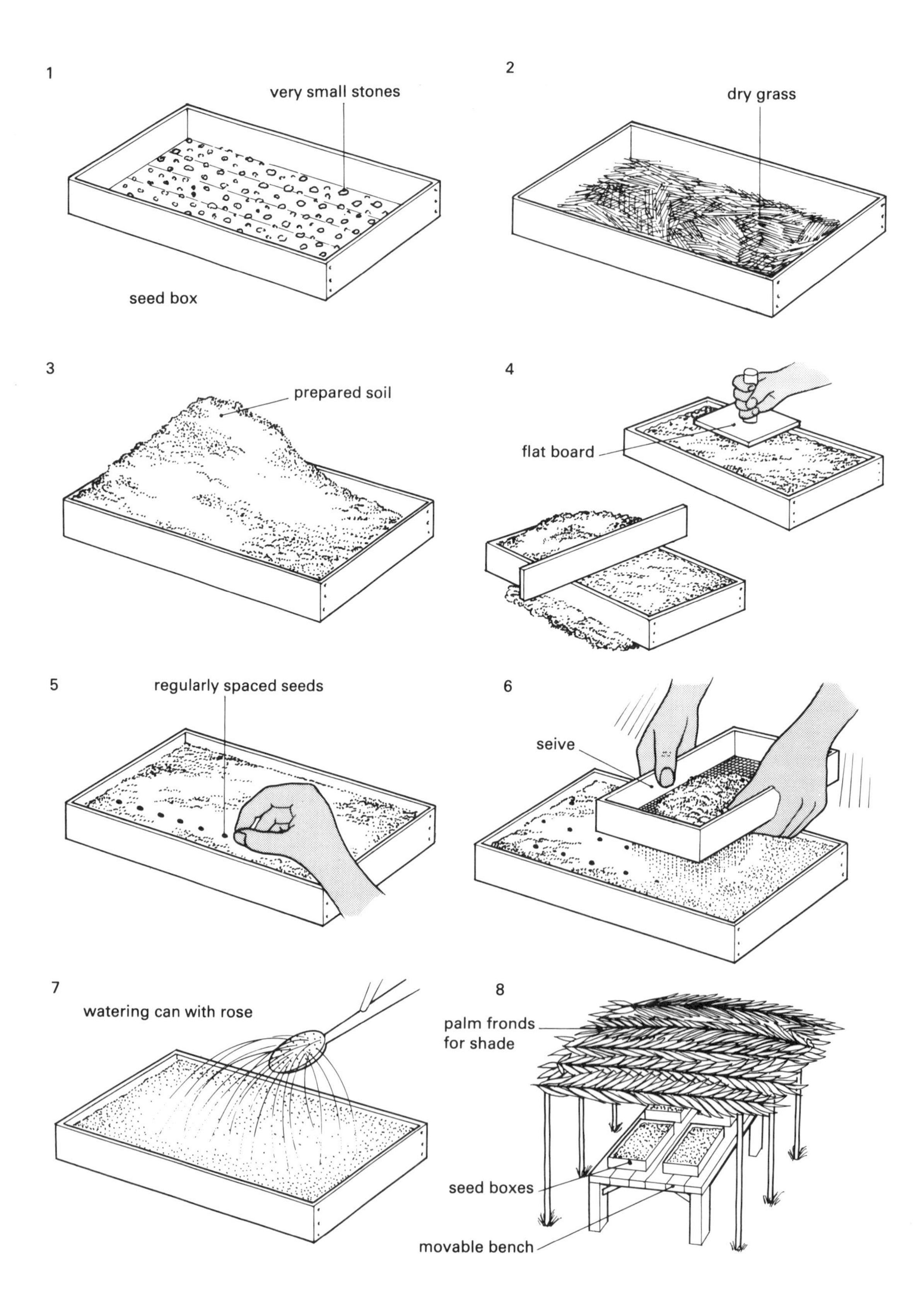
1
very small stones
seed box
2
dry grass
3
prepared soil
4
flat board
5
regularly spaced seeds
6
seive
7
watering can with rose
8
palm fronds
for shade
seed boxes
movable bench

Figure 8.7
Preparing a seed box and sowing seeds (opposite)

While any strong wooden box will do for the purpose, it is best that seed boxes are specially made; if they are the same size they can be handled with ease and stacked away carefully.

This is how to fill a seed box (Figure 8.7):

1 Put a layer of very small stones at the bottom.
2 Cover the stones with a layer of dry grass.
3 Prepare the soil. Mix about 50% of good dark topsoil with 25% chicken manure or well-decomposed compost and 25% sand. This should now be sieved or riddled into the boxes until they are within 3 cm of the top.
4 The soil should then be firmed down lightly into the boxes using a small, flat board.

This is how you sow seeds in a seed box:

5 Before sowing your seeds, read the instructions on the packet carefully. The seeds may be sprinkled over the surface of the soil or carefully spaced out; it will depend on the number of seeds and their size.
6 Sieve a little more soil over the seeds, about 1 cm is generally enough. Firm this soil down lightly using the small flat board.
7 Water the newly planted seeds with great care. Use a watering can with a rose. Start watering on the ground well to the side of the box then pass quickly over the seed box several times; in this way you do not wash away any soil or disturb the seeds.
8 Place the seed box in the protection of some shade. It is probably best if it is raised up off the ground away from animals and other pests.

Preparing the seedbed

If our seeds are to thrive then we must create the best possible conditions for them. The soil into which we plant the seeds (the seedbed) is most important.

A seedbed should be 'fine, firm and level'. If the soil is fine then the seeds will be in close contact with the soil, they will be able to absorb moisture and nutrients and will be able to develop good root systems. If the soil is firm, the soil will not sink or move once the seeds begin to grow. A level seedbed makes the work of weeding and watering much easier.

The seedbed should be dug very well, all weeds and roots should be removed. The soil should be rich and well manured; if you have compost available it should be dug into the soil. The compost will help to retain moisture and provide nutrients. The surface layers of the seedbed need to be prepared to suit the seeds being sown. Small seeds, such as lettuce and carrots, need a much finer seedbed than beans.

Figure 8.8
Palm fronds can provide shade for seedlings grown in a nursery

In some soils it is a mistake to make seedbeds too fine. When it rains we do not want our soils to form a hard surface cap (a layer of compacted soil) which will prevent air from entering the soil and will not allow water to drain freely away. This often happens with silty soils if too fine a seedbed has been prepared.

Shade for seedlings

Seedlings grown in a nursery require shade from the sun and protection from heavy rain. As the seedlings grow older and stronger the shade can be gradually removed.

Shading can be made from local materials. Usually palm fronds supported on sticks are quite adequate (Figure 8.8).

It is possible to buy special shade netting made of plastic. Although this is quite expensive, it is very effective. Shade netting is available in a number of forms which give different degrees of shade. Plastic shade netting is very effective if you wish to build a permanent shaded area or tunnel.

Seeds requiring shade during germination and the early stages of growth include lettuce, cabbage, cauliflower, pepper, onion and tomato.

ACTIVITY 5 Preparing a seedbed for beans

It is hoped that your school will have a garden in which you can grow some beans; if no land is available, perhaps you can grow the beans at home.

The area in which you grow the beans should be well fenced

against livestock; you cannot allow all your hard work to be destroyed by some stray animals!

If you are to plant a row of 40 beans, each person in the class will need a strip about 4 m long by 0.5 m wide. First the area may need to be cleared; the plot may need to be 'brushed' with a cutlass. Next the area must be dug to a depth of about 20 cm; you will probably need to do this with a fork.

If you have some compost or artificial fertiliser available it would be a good idea to mix some into the soil at this stage. Rather than digging it in over the whole plot you could mix it into the soil where you will plant your row of beans.

You will be pleased to learn that beans like a coarse seedbed; this means that you do not have to do much work!

Rake the plot until it is level. Do not worry if the soil is quite lumpy, the beans will not mind it at all!

ACTIVITY 6 Planting bean seeds

The time has come to plant your bean seeds!

You may find that bean seeds will germinate quicker if they are soaked in water for 12 hours before planting. If your school term is short this might be a good idea as the beans will mature faster.

Get your garden line and mark out a row in the middle of your plot. Cut yourself a stick 10 cm long; this is how far apart you

Figure 8.9 Planting bean seeds in the garden

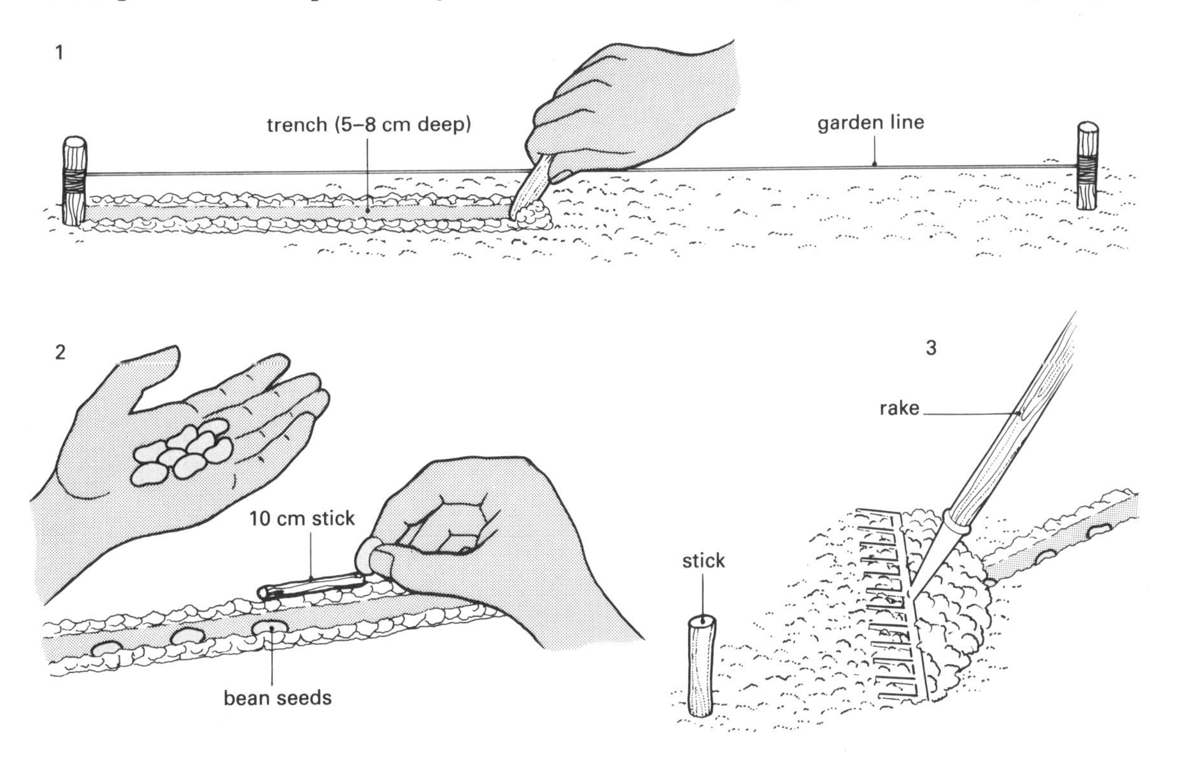

will plant the seeds. The seeds should be planted 5 cm to 8 cm deep either in holes, which can be made using a hand trowel or a dibber (a small hand tool used to make holes in the ground for planting seeds, bulbs or roots), or they may be placed in the bottom of a shallow trench (Figure 8.9).

Having planted your seeds you should carefully mark the ends of your row with sticks and gently cover over the seeds by raking soil over them. The soil should now be watered thoroughly.

Transplanting

Transplanting is a term used to describe the planting out of young plants from nursery beds or seed boxes into the field or garden.

Transplanting can take place when seedlings are big enough; this is usually at about 3 to 4 weeks. You should aim to transplant only strong vigorous plants.

The spacing of transplants will depend on the type of crop being grown and the size the plants will eventually grow to. For example, peppers should be planted in rows 75 cm apart with plants spaced at 75 cm within the rows (75 cm × 75 cm) and garden eggs or aubergines should be spaced at 90 cm × 90 cm.

For transplanting we usually use a hand trowel, and aim to move the plants with large balls of earth around their roots to protect them. Ideally, the balls of earth should be as big as your fist. After transplanting, firm the soil around each plant by hand

Figure 8.10 Transplanting seedlings

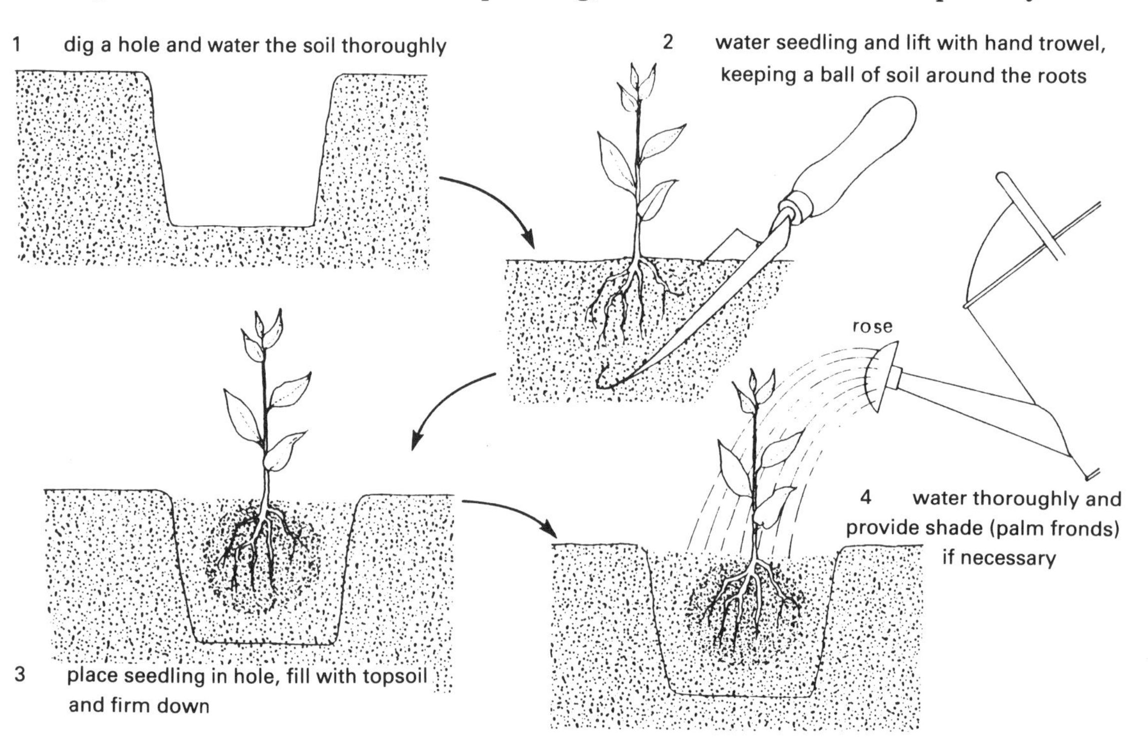

to prevent air pockets and to ensure good soil to root contact. Transplanting is best carried out in the morning or late afternoon but **not** when the sun is up! Transplanting during the heat of the day will result in the plants wilting and dying (Figure 8.10).

The newly transplanted plants must be watered thoroughly with a watering can or hosepipe. It may be a good idea to provide some shade for the plants until they are established. This shading may be moved progressively.

Caring for growing vegetables

Fertiliser application

If the soil is not very fertile you can increase your yields by applying some artificial fertiliser. We have seen that this may be added to the seedbed or to the growing crop.

Fertiliser should be applied to the soil around the vegetables after weeds have been removed; the crops should then be watered. For vegetables grown for their roots or fruits you might apply a compound fertiliser with an analysis of 10:10:10, at 8 g per square metre; this should be done only once or twice during the growing season. Sulphate of ammonia should only be applied to leafy vegetables at 8 g per square metre. Do not allow fertiliser to touch the stems or leaves of your vegetables as it may burn the plants.

Watering

Your vegetables should never be allowed to go short of water. Watering should be carried out in the morning or late evening; do not water when the sun is up as the leaves may 'burn' (turn brown).

Figure 8.11 Mulching and shading help prevent water loss in nurseries

Mulching

You may see many farmers putting vegetable matter around the base of their crops to protect the soil from the sunlight and to prevent the loss of moisture. This process is called mulching. Dried grass and rice straw are examples of materials that may be used as a mulch (Figure 8.11).

Pest and disease control

As a vegetable grower you will have to be very vigilant to stop pests and diseases attacking your crops! Every time you visit your crops, which should be at least once a day, you must be on the look out for signs of an attack.

As you weed or water your crops you should look at your plants carefully. Make sure that they all look healthy and that none are beginning to look weak. Turn over the leaves and look carefully at the undersides; do you see any signs of insects or the presence of a fungus (Figure 8.12).

If you think that your crops are being attacked you should seek advice. As you become experienced you will get to know many of the common pests and diseases and how to treat them. Some are easy to deal with others are more difficult!

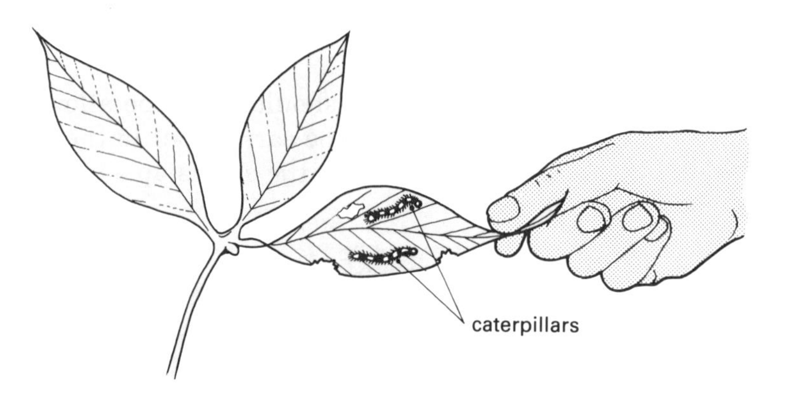

Figure 8.12 Regular crop inspection is essential

It is possible to pick off the larger insects by hand and kill them. Some can be scraped off using a small piece of wood, taking care not to damage the plants. Gently washing the leaves with soapy water may help to kill off insects.

Note: If you have to treat your crops with chemicals you must take great care. Follow EXACTLY the instructions supplied with the chemical. After treatment, do not harvest the vegetables until it is safe to do so!

ACTIVITY 5 Caring for your beans

During the growing period you are to care for your beans. This will mean visiting them every day if possible. Only by looking after them can you hope to have the best crop in your class! Once a week you should record in your exercise book how your beans are progressing; record the height that your beans have grown to and draw a picture of a typical plant.

As the seedlings grow, you should keep them free of weeds. By gently moving the soil around your row of beans you will prevent weeds from becoming established and you will prevent the soil from capping (forming a hard crust).

Unless there has been rainfall you should keep your beans well watered. Hopefully you will be able to borrow a watering can or bucket from the school. Be careful not to wash the plants out of the soil. You could apply a little mulch around the stems to preserve soil water; cut some grass for this purpose.

You can apply a little fertiliser to your beans if it is available. Your beans are leguminous plants so they do not need any nitrogen. Once the beans have grown to 15 cm you could apply 8 g per square metre of 0:10:10. Your plot would need 16 g in total.

You must keep a close watch on your beans for damage by pests and disease. You should use every opportunity to watch for an attack. Your teacher will help you identify the cause of any problem and help you to control it.

Harvesting vegetables

Despite the fact that there are many vegetables grown for a variety of plant parts, we can say that vegetables should be harvested when they are at the correct stage of growth. By harvesting vegetables when they are at their best we can enjoy them to the full.

If we are offering our vegetables for sale, we must present them in the best possible way. Only by doing so can we ask the highest prices. The marketing of agricultural produce is an area of activity which every farmer should be aware of. We will look at the marketing of produce in a later book.

ACTIVITY 7 Harvesting your beans

This is the most rewarding part of the whole exercise! When your beans are ready for harvest you may pick them and weigh them. Record the weight in your exercise book. Who grew the most in your class?

ACTIVITY 8 Marketing of vegetables

Visit your local market every week for a whole year (or term). Note down in your exercise book the vegetables on sale. Record the prices being charged for each of them. You will see that the prices vary as the season for each vegetable starts and finishes. Why is this? Make notes on the quality of the vegetables on offer? How do the traders make them look more attractive? By recording what is on sale at the local market over a period of time you are creating a calendar which reflects the farming year!

Some key words and terms in this chapter

Vegetable Any one of various herbaceous plants having parts that are used for food.

Germination The sprouting and growth of seeds.

Damping off The dying of young seedlings due to fungal attack.

Nursery An area of land set aside for the raising of young plants.

Mulching The placing of material on the soil around the stems of plants to prevent water loss. Dried grass may be used as a mulch.

Transplanting The process of planting a young plant from a nursery or seed box into the field or garden.

Thinning Removing excess plants. This allows the remaining plants to develop to their full potential.

Shade netting Commercially available sheets of plastic net which provide varying degrees of shade to young plants.

Seedbed A place in which seeds are sown.

Capping The formation of a hard surface on the top of the soil by rain. Silty soils are very prone to capping.

Dibber A simple hand tool used for making holes in the ground to plant seeds, bulbs, roots and transplants.

Exercises

Multiple choice questions

Write the correct answers in your exercise book.

1 Which of the following materials is **not** used in the preparation of seed boxes for the sowing of seeds?
   a stones
   b dry grass
   c sulphate of ammonia
   d good topsoil

2 Which of the following would we **not** normally consider a vegetable?
   a spinach
   b cabbage
   c garden egg
   d rice

3 Which of the following vegetables is **not** raised in a nursery before being transplanted to the field?
   a lettuce
   b carrot
   c garden egg
   d tomato

4 Removing seedlings from seed boxes or nursery beds with a ball of earth using a hand trowel is called
   a shading
   b thinning
   c weeding
   d transplanting

5 Which of the following tools is used for making holes in the soil to plant seeds, bulbs, roots or transplants?
   a fork
   b dibber
   c hoe
   d rake

6 Which of the following is used to maintain moisture around vegetable seedlings?
   a mulch
   b fertiliser
   c capping
   d pesticide

Missing words

7 Copy the following passage into your exercise book and fill the gaps using the words below.

Only by using the best ...... can a farmer hope to be able to grow the best ...... These should be bought from a recognised .... The farmer should read the ...... on the packet and follow them closely. The crop should be looked after very carefully. When the crop is ...... it should be ...... If it is to be ...... the farmer must make sure that the ...... of the goods is as high as possible. Only by selling first class vegetables can a farmer hope to make a large ......

vegetables	**profit**
marketed	**seeds**
mature	**agricultural merchant**
instructions	**harvested**
quality	

Crossword puzzle

8 Copy the diagram into your exercise book and complete the following clues.

Across

1 Placed in the bottom of a seed box, they assist in drainage (6)
2 Occurs when seedlings are attacked by a fungus (7,3)
3 Used to establish many vegetable crops (4)
4 Material used to conserve moisture around growing crops (5)
5 Often needed as a daily operation in the growing of vegetables (8)
6 A vegetable grown for its tasty red fruits (6)

Down

7 The planting out of young plants into a field or garden (13)
8 Seeds do this when they begin to grow (9)
9 A vegetable grown for its leaves which form a tight green ball (7)
10 These are responsible for a lot of crop damage (5)

Points for discussion

9 How can you tell when vegetables are ready for harvesting? If vegetables have to be sent to the local town, how can you make sure they arrive at the market in good condition?
10 How can farmers improve the marketing of their vegetables and make more money?
11 Why are vegetables so expensive to buy in the market?
12 Were you successful in growing your beans? If so, to what do you attribute your success? If not, discuss the reasons for any failures and make notes in your books on the ways in which the problems can be overcome in the future.

CHAPTER 9

The livestock industry in Sierra Leone

Introduction

In this chapter we will look at the importance of farm animals to the people of Sierra Leone. We will discover the uses to which the animals are put. We will compare the modern, scientific ways of keeping livestock with our traditional systems. We will learn that our livestock must not be allowed to compete with our wildlife and that we must control the hunting and killing of our wild animals. We will look at different classes of livestock, depending of the way they digest (break down and absorb) their food. Towards the end of the chapter we will look in more detail at the keeping of livestock in the traditional way found in most of our villages.

What do we mean by livestock

Livestock is a term used to describe animals (cattle, sheep, goats, horses etc.) that are kept for domestic use. Livestock are usually kept on a farm or are looked after by herdsmen as they move through the countryside (Figure 9.1). Domestic pets (dogs, cats etc.) are not usually included in a definition of livestock.

You will often hear people talking about 'livestock husbandry'; this means the skill or science involved with the caring and management of farm animals.

The domestication of livestock

In previous chapters we looked at the early development of agriculture and the way in which crop production may have begun. Let us now try and imagine how our livestock industry may have developed.

When hunting for meat, early people would have killed mature animals and may have brought back the motherless young to the village. These immature animals would have been allowed to grow before being eaten. The people of the village would have cared

Figure 9.1
Livestock is a term which includes all animals kept for domestic use

for them and found that they became quite tame. They would have found that it was easier and more reliable to use these animals than to go off hunting each time meat was required. By allowing the animals to breed, early farmers were assured of a steady supply of meat! People would have found that these animals could also be used to supply other products, such as milk, wool and hides.

Over a considerable period of time, several types of animals supplying meat, milk, hides, skins, wool or hair, and transport were domesticated. In this way people would have built up herds or flocks of animals. By picking out the best animals for breeding, farmers would have developed special varieties or breeds of livestock. Slowly these breeds would have become less and less like their wild relatives.

Having built up the number of animals in their care, early people had to develop ways of looking after them. The animals had to be fed and protected from predators (animals that prey on others). The breeding of the animals had to be controlled and the animals treated for pests and diseases, and so the skills of animal husbandry gradually developed.

ACTIVITY 1 Livestock and people

In this activity you have to think of all the types of livestock kept by farmers. You are to say what the collective (group) name for the animals is and for what purpose the animals are kept.

Copy down Table 9.1 into your exercise book and complete the columns. (To help you to start, chickens have been used as an example.) When you have finished your teacher will draw together all the information from your class. You will see how important livestock are to us!

Table 9.1 Livestock and their uses (Activity 1)

Livestock	Group name	Purpose and/or product(s)
Chickens	Flock	Eggs, meat, feathers (cushions, pillows and decoration), manure

Livestock production and wildlife

We are lucky when compared to early people! Our farmers are able to give us a steady supply of food and it is becoming less important for us to hunt wild animals.

Today, the human population in Sierra Leone is quite high and it is rising all the time; this is putting pressure on our countryside and all our natural resources. This is especially true in the case of our wild animals, and the forests and wild places in which they live.

Modern hunters use guns and rifles that are much more efficient than the spears and clubs used in previous centuries. They are able to kill a great number of animals very easily. The number of our wild animals is decreasing rapidly and there is a very real danger that, by the time you become an adult, many of our special animals will have become extinct. We cannot allow this to happen! We must preserve the balance of nature and ensure that all our wildlife thrives.

A limited amount of hunting may still have to take place to control any large numbers of animals that may attack our crops. However, we can no longer afford to hunt or set traps for our animals and kill them for food.

We must increase the production from our domestic livestock to feed our growing population. In increasing our production, we cannot allow our farmers to destroy more and more bush and forest; these areas must be protected. We must learn to make better use of the land already under our care. We cannot allow erosion (the removal of soil by water and wind) and neglect to spoil our land, forcing our farmers to move on to new areas.

The care and preservation of our natural heritage is something which must concern us all.

Livestock terms

Before we can learn about livestock we should understand the terms farmers and agriculturalists use to describe animals of different types and ages. Many of these terms will be known to you; others may be new.

ACTIVITY 2

Livestock terms

Let us discover what these terms are. Copy Table 9.2 into your exercise book and try to fill in the gaps. Your teacher, with the help of your class, will help you fill in any gaps you may have left. (Note: It may not be possible to fill in all the spaces in the table, and some of the spaces may have more than one answer!)

Table 9.2 Livestock terms (Activity 2)

Type of stock	Adult male	Adult female	Young	Group of young
Cattle				
Sheep				
Goats				
Pigs				
Horses				
Rabbits				
Chickens				
Ducks				

Livestock production in Sierra Leone

We have seen that the livestock industry is very important to Sierra Leone. Let us look at a map of the country and see where the main centres of production are located (Figure 9.2).

You will see that most of the cattle, sheep, and goats are raised in the north of the country, on the interior plateau and mountains. In this region there is plenty of grassland on the lower slopes which is most suitable for livestock or pastoral farming. We have already talked about the Fulani herdsmen who travel with their herds through the semi-arid regions (Figure 9.3). These people make an important contribution to the agricultural production of our country.

Very few animals are raised in the south and east of the

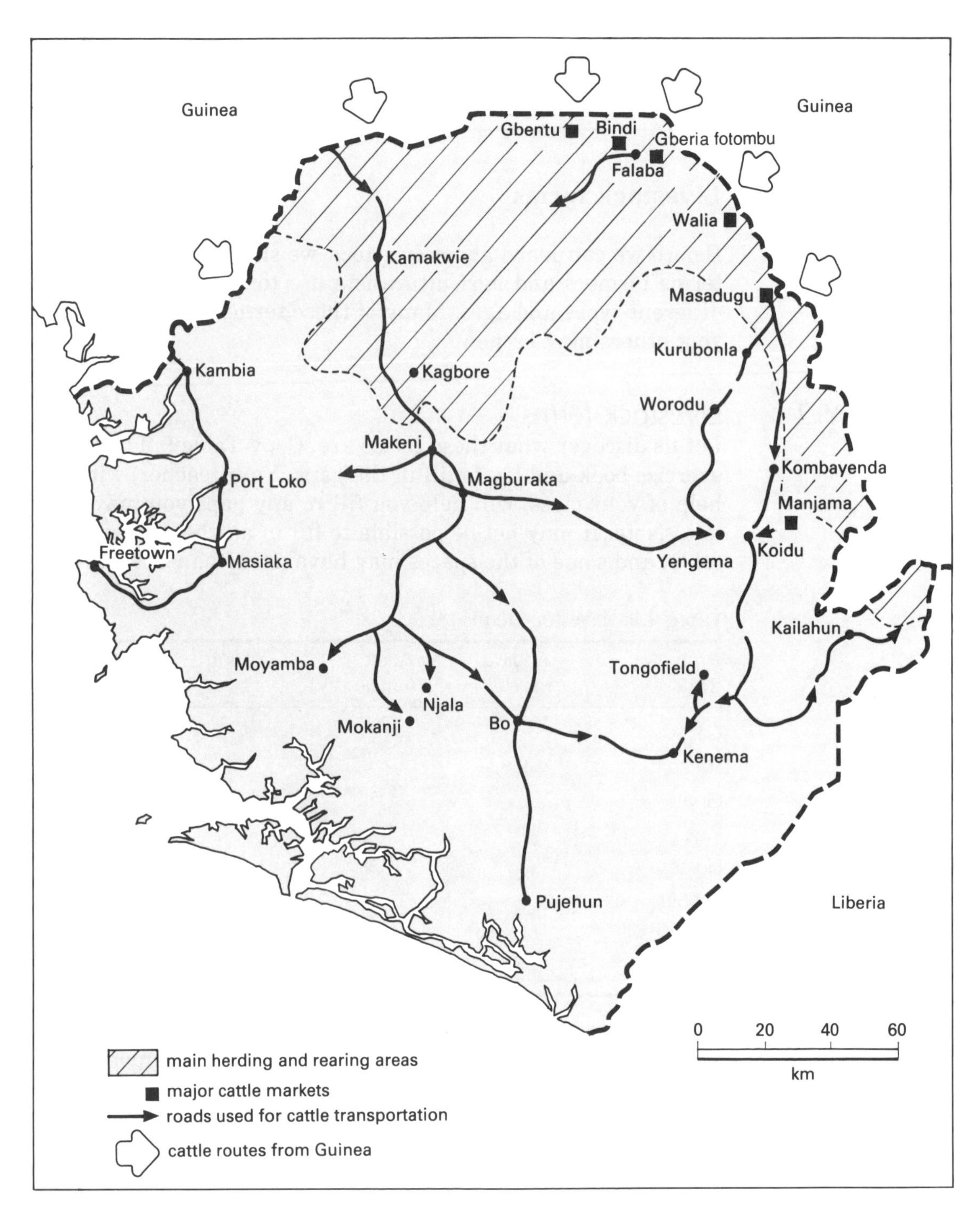

Figure 9.2 Main areas of livestock production in Sierra Leone

country. Here the climate is wetter and more humid; this encourages the tsetse fly to breed. The tsetse fly causes na'gana (sleeping sickness) in cattle, and large areas of our country are not really suitable for cattle production.

You will notice from the map that many of the large-scale poultry and pig enterprises can be found close to centres of

Figure 9.3 Fulani cattle survive well in semi-arid regions

population. For example, poultry and pigs are produced at Doray's Poultry Enterprise, Freetown, the Alton Poultry Division at Masorie, the Teko Livestock Division in Makeni, the Poultry and Piggery Centre at Njala University College in Southern Province and at Rokel Animal Farm near Freetown. Why do you think these farms are found here?

ACTIVITY 3 Livestock production in your area

Study the map shown in Figure 9.2. Find where you live. Is the map correct in telling you the main livestock types that are kept in your area? Is a large livestock production unit to be found in your district? What is it called? What is produced there? Where are the products sold? Who owns and manages the unit? Ask your teacher to take you to visit it.

Have you or your classmates visited other areas of the country? Were any livestock kept there? Describe what you saw.

The purposes of livestock in our economy

If someone was to ask us "Why do people keep livestock?" we would probably say that it was for food, or transport! This is essentially correct, but is only part of the story! Livestock play other roles in our economy and society; let us remind ourselves about them:

- We have already established that livestock provide us with food (meat, eggs, milk) and other products (wool, leather) used in our everyday lives.

- Livestock are an important source of income; this is true whether the livestock are kept on a very large farm or are owned by a small farmer in a village. In our traditional agriculture, animals may be the only source of income for a very poor farmer.
- Animals may be kept as a form of insurance. In times of emergency or need, our traditional farmers may sell some livestock to raise money. Animals may be sold if there is a disaster or the farmer has to meet some social obligation (wedding or funeral). Livestock owners may consider that their wealth is safer walking about than being kept as cash in their houses or in the bank! The animals may also breed and increase the owner's wealth!
- Livestock are an important source of employment for men, women and children. A herd of cattle may need several people to look after them.
- Livestock are an appropriate and reliable source of power; we learnt about this in Chapter 5. Farmers can use animals to transport goods or to work the land. We have seen that animal power should become more important in Sierra Leone.
- Livestock give their owners social status. A person with large numbers of livestock may be highly respected. This is especially true in the north of the country where large herds of animals are kept.
- Livestock often play a part in our religious and cultural ceremonies. Can you think of occasions when livestock are used for this purpose?

Livestock products and their uses

Let us look in some detail at the main products that livestock provide us with.

Milk

Cows are our main source of milk. Goats and sheep are rarely milked in Sierra Leone; flocks of these animals are less common and the animals give considerably less milk than cows.

Milk from cows, goats and sheep is a valuable food for both children and adults as it provides proteins, carbohydrates and minerals. Milk can be drunk fresh. If milk is not to be used immediately it should be kept in a refrigerator or it will quickly turn sour.

Milk may be processed into a variety of products (Figure 9.4). A place where milk is processed is known as a dairy and the products are called 'dairy products' or 'milk products'. Cheese and yoghurt are two dairy products. Can you name any others?

Figure 9.4
Dairy products

Do not forget that we import a number of dairy products into our country. What are these? Where do you think they come from? If we were to develop our own dairy industry we would not need to import so many of these products. Do you know of any dairies in Sierra Leone?

The first milk that a cow gives after its calf is born is very rich; it is called colostrum. The calf needs to drink this milk as it provides protection from disease. A good farmer will make sure that the calf receives its colostrum during the first few days of life. The farmer should not sell it or give it to the family.

Meat

Meat provides us with essential proteins that are needed to build our bodies. Proteins are an essential part of our daily diet. Farm livestock provide us with much of the meat we eat (Figure 9.5).

Livestock are usually slaughtered (killed) in the village or in a special slaughterhouse near a town. Meat, like milk, will quickly go bad unless it is used immediately, refrigerated or frozen. Freezing is the best way to preserve meat until it can be used; keeping meat fresh in this way is expensive so you must expect that it will cost a little more.

What sort of meat is sold in your village? How much does it cost? Record in your exercise book the prices of different meats? Is all the meat from the same animal sold at the same price or are different parts sold at different prices? How is meat processed locally? Where do the animals that are slaughtered come from?

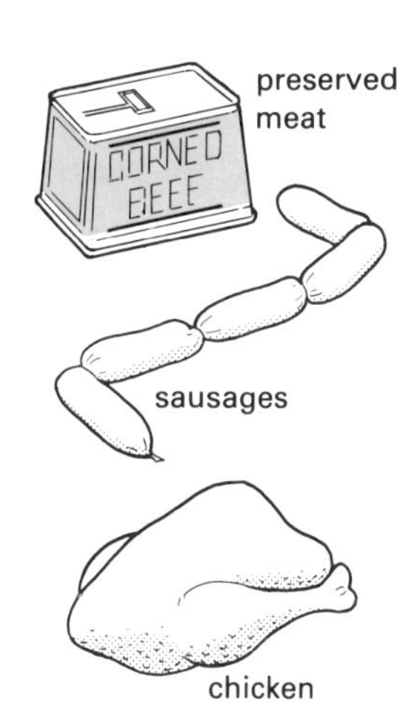

Eggs

Eggs from poultry are another important livestock product providing us with protein. Eggs, whether they are produced by local farmers or in large intensive poultry farms, are always in great demand.

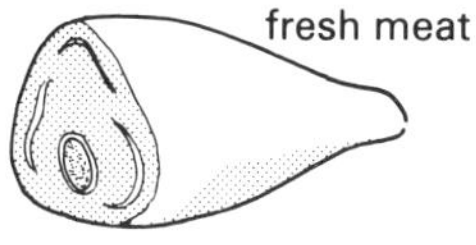

Figure 9.5
Meat products

Hides and skins

Hides and skins are valuable as they can be cured (processed) into leather; this is often carried out in villages. The thick hides from cattle are made into shoes, and the thinner skins from sheep and goats are made into gloves and other light goods. The very best hides and skins are exported (Figure 9.6).

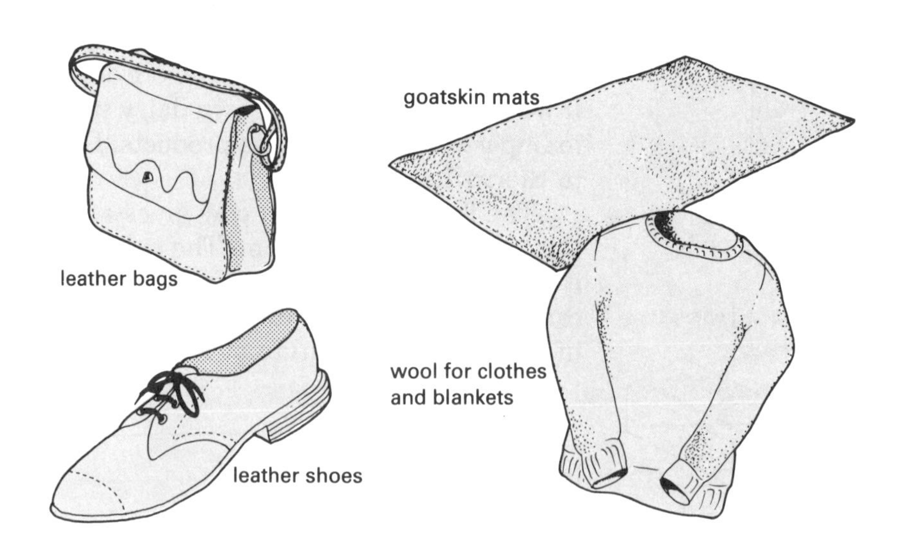

Figure 9.6
Products made from hides and skins

Wool and hair

In Sierra Leone, wool and hair from sheep and goats are not very important to our farmers. The breeds of animals found in the tropics do not need thick coats to keep warm. Our sheep have a mixture of hair and short wool to keep cool. There are breeds of goats in the drier regions of the country that grow a very fine hair which is in demand.

Slaughterhouse waste

Waste meat, offal and blood are taken from slaughterhouses to factories where they are processed into a valuable protein feed for livestock.

Bones are dried, ground up and combined with other materials for sale as organic manures. Fertilisers based on bone meal are very effective.

Manures

Livestock faeces, when mixed with straw or other vegetation, form manure which is an excellent, cheap and readily available fertiliser (Figure 9.7). Manure must be quickly buried in the soil to keep it moist so that it rots down and does not lose the nutrients contained in it. Manure left on the soil surface dries up and is of little value to the farmer.

You will now appreciate how important livestock are to our farmers and to the country! It is important that our farmers look

Figure 9.7 Manure from intensive animal production makes a valuable fertiliser

after their animals as well as possible. In this way the animals will provide a good financial return and the country will benefit from the many products they provide.

ACTIVITY 4 Local livestock count

Copy Table 9.3 into your exercise book and label the columns as shown.

In this exercise you will visit five farmers and you will find out what livestock they own and the use to which the animals are put. If your teacher tells each group in the class to talk to different people you can get quite a lot of information about livestock in your village. When you return to the classroom you will be able to work out the average numbers of each type of animal kept by farmers.

Table 9.3 Local livestock count (Activity 4)

Name of farmer	Type of livestock	Number	Uses
(Example only)			
A Kabba	Cattle – milk	6	Milk and beef
	– beef	4	Beef
	– calves	3	
	– oxen	2	Work
	Poultry – chickens	12	Eggs and meat
	– ducks	10	Eggs and meat
	Rabbits – does	6	Breeding and meat
	– bucks	1	Breeding

Classification of livestock

We can place farm animals into groups according to their methods of digesting food.

Ruminant digestive system

We know that cows, sheep and goats eat large quantities of bulky food, such as grass, the leaves of trees and bushes, and the dry residues left in the fields after harvest. This bulky food contains a lot of a very tough and indigestible carbohydrate called cellulose. Before these animals can use cellulose they must break it down. To do this they have developed very special digestive systems.

The food that ruminants eat is so tough it needs to be chewed twice! The food is first chewed as the animal takes in its food. It is swallowed and then brought up from the stomach to be chewed again. If you watch a ruminant at rest you will probably see it chewing. Watch its throat carefully! You will see it pass a ball of grass up into its mouth, chew it and then swallow it again! Another ball of food is then brought up into the mouth. This is called 'chewing the cud'.

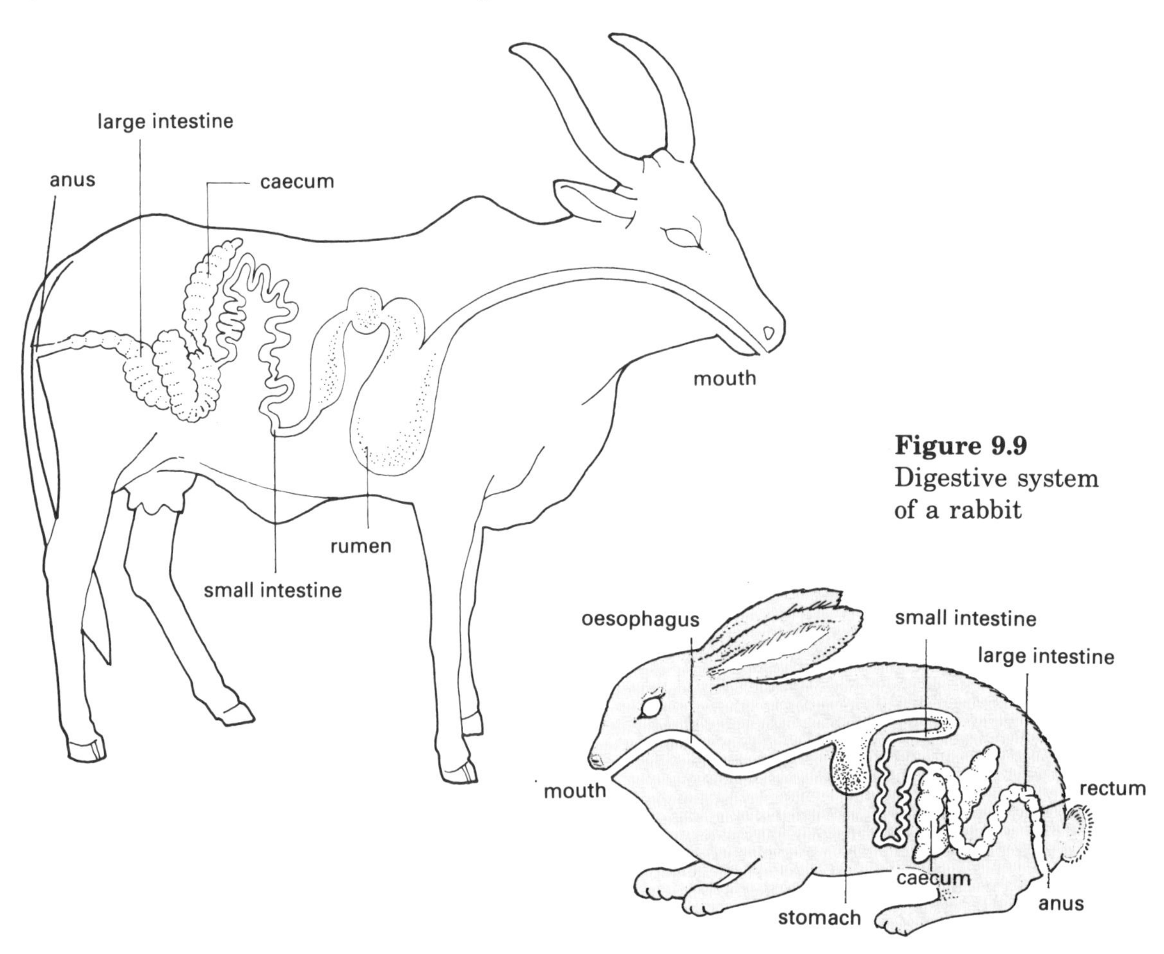

Figure 9.8 Ruminant digestive system

Figure 9.9 Digestive system of a rabbit

The main part of the digestive system shown in Figure 9.8 is called the rumen. You may think of the rumen as a large tank, containing water and millions of bacteria, which churns the food about and mixes it up! The bacteria in the tank are able to break down this tough cellulose into a form that the animals can absorb. During this process, which is called rumination, methane gas is produced. Carefully watch a cow eating and you may see (and hear) her belching up this gas! The partly broken down food is then passed into the rest of the digestive system.

Rabbit and horse digestive system

While rabbits and horses eat the same kind of bulky food as ruminants they do not have a rumen. They have developed their own method of breaking down cellulose.

In Figure 9.9 you will notice that these animals have an enlarged caecum, which breaks down roughage in much the same way as the rumen. However, food is not passed back to the mouth for rechewing. The horse does not digest its food very well and needs to eat large quantities of food to compensate. The rabbit, on the other hand, manages to make better use of its food by eating some of its droppings!

Pig digestive system

Pigs have the same type of digestive system as humans. Such animals are said to have 'simple stomachs'.

Pigs do not eat large quantities of bulky foods but prefer a more concentrated diet. In the wild, pigs will eat a wide variety of roots, grains, fruits and nuts. If we keep them under intensive conditions we feed them a concentrated diet of cereals and protein meal. The food is in a much more digestible form and does not need a complicated digestive system. The food is broken down in the stomach and then absorbed in the intestines (Figure 9.10).

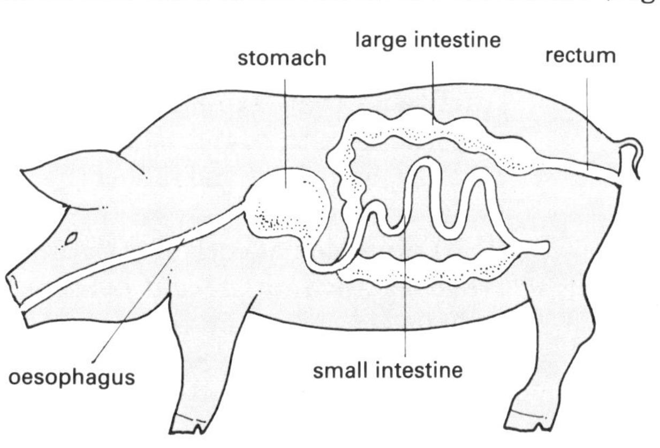

Figure 9.10
Digestive system of a pig

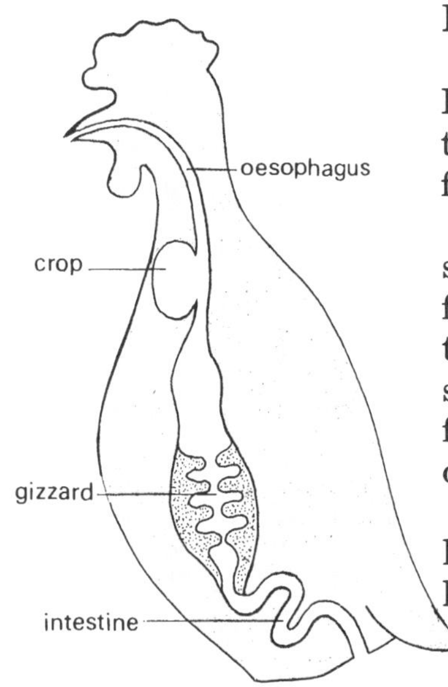

Figure 9.11
Digestive system of a chicken

Poultry digestive system

Poultry do not have teeth and cannot grind down their food in the same way as cattle or pigs. They have developed their own form of digestion.

As poultry are unable to chew food in their mouths they must swallow it whole. From Figure 9.11 you will see that the food is first held in a storage pouch called a crop. It is then passed to the gizzard. This organ contains grit (very small fragments of stone) and has a rough leathery surface; it is able to grind the food up finely. The food may then be digested in the remainder of the bird's gut.

Poultry must eat grit to break down their food. If poultry are kept intensively you will have to supply them with it. Poultry kept out of doors will be able to find their own grit.

Water

All animals need water to digest their food and to stay alive. When you are looking after your animals, make sure they always have a plentiful supply of clean, fresh water. The water you give to your livestock should be so pure that you would be able to drink it yourself!

Extensive and intensive livestock production

'Extensive' and 'intensive' are agricultural terms we are already familiar with!

You will already understand that 'extensive' refers to the production of crops or livestock using large areas of land and few resources. Our traditional agriculture falls into this group. In extensive agriculture we would expect that the yields from crops and livestock would be fairly low.

'Intensive' refers to the use of a lot of resources to produce a high output of crops or animals using a relatively small area of land. The farming you might see on Government farms, and commercial estates and plantations, is of this type.

Do not think that intensive agriculture is always more profitable than extensive agriculture! Unless a farmer is very careful there is a danger that more money may be spent on inputs than is made from the sale of the produce; this would mean a loss would be made.

Let us look at each of these systems in turn and consider their relative advantages and disadvantages. You will see that the advantages of one system can be the disadvantages of the other! Perhaps, before you read the following section, you could try and think of some advantages and disadvantages yourself.

Extensive systems of livestock husbandry

We must not think that there is anything inferior or second class about extensive agriculture when compared to intensive farming. People practise extensive agriculture in response to certain circumstances. It is often the most appropriate form of agriculture to carry out in a given region. This is particularly the case in the arid lands of Africa where large areas of land could not support intensive agriculture because of the poor soil and lack of water. It is also appropriate for poorer farmers who are unable to afford buildings, equipment or expensive breeds of animals (Figure 9.12).

Some of the advantages of the extensive system of keeping livestock are listed below:

- There is no need to make a large investment of money in buildings and machinery. The livestock may move over large areas and such facilities would be unnecessary.
- Several livestock owners can pool their resources. In this way one herd or flock is formed under the control of one or two herdsmen or shepherds.
- Livestock kept under extensive systems are able to move about freely. They become hardy and tolerant to a number of pests and diseases.
- Because they fend for themselves, they generally have a balanced diet and do not suffer nutrient deficiencies.
- The herdsman may be able to charge the owner of the land on

Figure 9.12
Extensive livestock keeping suits the prevailing conditions in Africa very well

which the stock spend the night for the manure left on the ground.

The disadvantages of the extensive system are:

- The stock may have to move large distances to find food and water. The dry season is often a very difficult time and the animals may suffer. As the livestock walk they use up valuable energy and lose condition.
- The stock have to be tied or kraaled (kept in a compound) at night, to protect them from wild animals and thieves. This involves a lot of labour.
- The herdsmen have to stay with the stock at all times, and may have to move with them far from home.
- It is difficult to control breeding. The herdsmen may not be able to select the best animals to improve the herd or flock.
- As the animals are on the move, they are not easily handled; pests and diseases cannot be dealt with effectively.
- The animals are at considerable risk. They may be at risk from thieves, wild animals including snakes, and traffic accidents.

Intensive livestock husbandry

Intensive livestock farming is practised in many developed countries where there are large human populations and where there is not a lot of land. The livestock are kept in concentrated numbers and may be housed in special buildings for most, if not all, of their lives (Figure 9.13).

Some of the advantages of intensive livestock production are:

- The health of the animals can be monitored easily and pests and diseases can be controlled.
- Environmental conditions can be automatically regulated ensuring that the animals are comfortable at all times. Ventilation (air movement) and temperature (heat and cold) are often controlled.
- The feeding of the animals is more accurate; each animal will receive a balanced diet of high quality food. Feeding may also be automated (carried out by machine).
- As the livestock do not have to search for their food, they will produce more. The animals may put on more weight or produce more eggs.
- It is possible to produce large quantities of high quality produce. Traders and consumers like to have goods that are of a standard type and quality. Large groups of animals can be grown to the required size and weight and can be sold as one unit.
- It is possible to control breeding. By keeping males and females apart, selected animals can be used for breeding to improve the herd or flock.
- The animals are more easily protected.

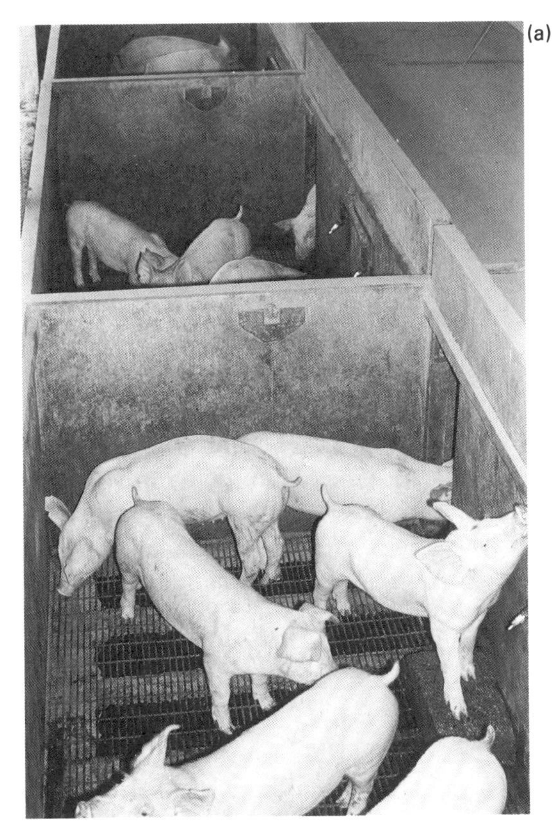

Figure 9.13 Intensive livestock production:
(a) pigs
(b) chickens
(c) cattle.
This type of production requires a lot of money and technical knowledge if it is to be successful

Intensive livestock production has several disadvantages. These include:

- A lot of capital (money) is required to provide the livestock, buildings and equipment needed for this type of farming.
- Those who run the system must be skilled and vigilant.
- Veterinary costs may be high.
- Pests and diseases may spread quickly when so many animals live closely together.
- Unless the system is well managed the animals may suffer stress (fear and discomfort); this can cause loss of production. Intensive livestock production requires a high level of technical knowledge.
- Large quantities of high quality feed and water must be available to the stock. If there is a breakdown in the feed or water supply there must be some form of back-up to remedy the problem, otherwise the livestock will die.
- Farmers must have a reliable market for the produce and they may need transport to bring their animals or produce to market.
- The running costs of an intensive livestock unit are high. In particular, the systems require a lot of labour to carry out the

routine tasks (feeding, watering, etc).

- Leg weakness occurs because the movement of animals is restricted.

ACTIVITY 5 Visit to an intensive livestock unit

Your teacher will arrange for you to visit one or more intensive livestock units in your district. (Remember, these units do not have to be very large to be intensive; a small rabbit unit can be very intensive.)

Try to find out as much about the unit as possible. Write plenty of notes in your notebook. Ask the farmer about the type and breeds of animals that are kept. What are the animals kept for? Where is the produce sold? What are the animals fed on? Where does the feed come from? How are the animals looked after? What problems are found in keeping the animals intensively? Are periodic visits made by a veterinary doctor or assistant to treat the animals?

Traditional livestock production

Let us now look at livestock production as it is carried out in the countryside and villages of our country. This small-scale production is often overlooked as it is not considered important. However, it makes a great contribution to our economy. Try to imagine all the people who are fed from the animals that are kept in the countryside.

You will see that most people keep their animals extensively. How and why do you think this system suits the people so well?

Poultry

You will remember that the term 'poultry' includes chickens, ducks and geese and any birds that are kept by farmers for food or eggs.

Chickens

The keeping of chickens is a tradition in the villages of Sierra Leone where they are raised for ritual sacrifices and special occasions, such as marriage ceremonies or religious feasts.

The birds are kept in a free-range system (an extensive system where they are able to move about freely in the village or compound). Production is low and haphazard; the eggs and meat are of variable quality. The birds pick up whatever food they can find. Occasionally, they may be fed some grain or household waste by their owners. The chickens grow slowly. At night they 'roost' or sleep near the owner's house; they may roost in trees, in the kitchens, in abandoned houses or in small cages or coops. The chicken droppings are collected by the farmer and used as manure.

There is little attempt made to control the breeding of chickens or to improve the flocks. The chickens make their nests and lay their eggs in a wide variety of places. If the owner can find the nests, the eggs make a good addition to the family diet. If the nests are not found, the hen may sit on the eggs and hatch them.

The chickens are exposed to many dangers; they may be killed by hawks, snakes or vehicles, or may be stolen. Diseases can kill a whole flock.

Ducks

At village level, ducks are kept in much the same way as chickens; that is, they are kept in an extensive system and little attention is paid to their husbandry. Ducks are strong birds and their habits make them ideally suited to being kept in this way. They are able to find their food in the fields, and are less likely than chickens to catch diseases.

The most common breed in West Africa is the Muscovy duck, but the small Khaki Campbell breed is found in many villages. Ducks lay fewer eggs than chickens. They do not make good mothers and the task of sitting on duck eggs is often given to a hen!

Many people do not eat ducks as they find the meat unpalatable; this may explain why ducks are not more popular. When ducks are slaughtered, they may be hung to allow the blood to drain away before they are cooked.

Many people hold superstitious beliefs about ducks. For example, people believe that ducks have got supernatural powers and that, if a duck is killed by a vehicle, the driver is sure to have an accident later! That is why, when a duck and her ducklings are crossing a street, drivers wait for them! Do you believe that? What other beliefs about ducks have you?

Pigs

At village level, pigs are also kept in an extensive system. A few pigs are often to be found in each village where they act as scavengers, searching for household waste or fallen fruit, and grubbing for roots. As their diet is poor these pigs grow slowly (Figure 9.14).

The standard of pig husbandry at village level is low and little attempt is made to control the breeding of the pigs or to care for the young piglets, so only the strongest survive. However, pigs provide meat and skins for their owners.

Pigs are not popular domestic animals in Sierra Leone for social and religious reasons. Pigs, being simple-stomached animals, are very similar to humans and some of their parasites may be passed to people. Those of the Muslim faith are forbidden

Figure 9.14
A local breed of pigs in Sierra Leone

to eat pig meat. Pigs can be fed on scrap food, that is leftover human food, generally called swill.

Cattle

The Fulani or Fula people are the traditional cattle owners of West Africa and their herds are to be found in the semi-arid uplands of Sierra Leone. Cattle in the wetter regions of our country are fewer in number and, while they are kept for milk, manure, hides and meat, they are also used as work animals.

If cattle are being raised for meat production they do quite well in an extensive system. If milk is to be produced, it is best that the animals are kept in an intensive system where they are close to the dairy.

Most of the cattle reared in Sierra Leone are of the N'dama breed (Figure 9.15); these were originally brought to Sierra Leone from Guinea by the Fulas. N'dama cattle are very resistant to sleeping sickness (caused by the tsetse fly), which can affect cattle of other breeds. The N'dama are small (about 180 kg) when fully grown and are kept for meat and work rather than milk production; they are considered to be a beef breed rather than a dairy breed.

Most of the cattle reared in Sierra Leone are to be found in the Bombali and Koinadugu districts. This is due to the fact that:

- These districts have the largest areas of savanna (grassland) in Sierra Leone and are most suitable for cattle rearing.
- They have a dry climate, which discourages the spread of sleeping sickness.

Figure 9.15
N'dama cattle are well suited to the tropical climate of Sierra Leone

Sheep and goats

Sheep and goat husbandry is carried out in much the same way as cattle production.

The northern areas of the country, with savanna grasslands, allow farmers to keep larger flocks or herds, which are looked after by shepherds or goatherds.

In the remainder of the country, farmers tend to own far fewer animals. The sheep and goats are allowed to browse or graze on what feed is available. It is desirable that these animals should be tethered; this will prevent them from straying into other people's gardens where they may cause a great deal of damage (Figure 9.16).

While goats are very useful animals, they can also be very destructive. Goats are very hardy and resistant to disease. They

Figure 9.16
Dwarf West African goats

Figure 9.17
A fish pond in a rice paddy in Sierra Leone

can survive on a wide range of plants and will readily eat the leaves of trees and shrubs.

However, if goats are allowed to become too numerous they can damage the environment. This is happening in many parts of Africa, especially in the arid and semi-arid areas where the environment is very delicate. Here the goats are stripping the leaves from the trees and shrubs and eating seedlings; this is causing the deserts to grow larger and larger.

Fish

Fish are an important source of high-quality protein in the diets of many villagers in Sierra Leone. While fish farming is something relatively new to our country, people have traditionally caught or trapped fish from the sea, rivers, streams and swamps.

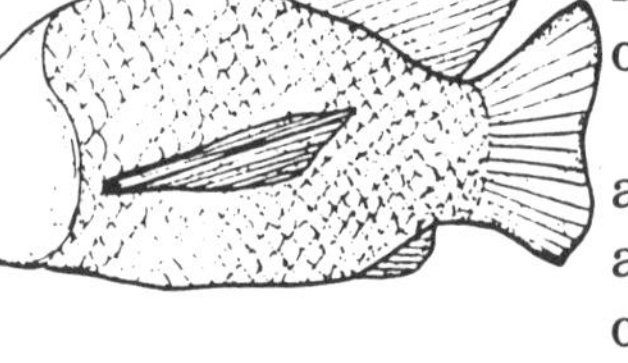

Figure 9.18
Tilapia are commonly used for stocking small fish ponds in the tropics

In fish farming, a stream is diverted into a pond or reservoir and stocked with a suitable breed of fish (Figure 9.17). Tilapia are commonly used for this purpose. The fish in the pond will feed on the natural pond life and may also be fed with additional food by the farmer. Waste cereals, such as maize, sorghum and rice offal make good food for tilapia. When the fish are big enough, the pond is netted and the fish are harvested. Tilapia are very suitable for farming as they reproduce easily, grow quickly and soon build up numbers (Figure 9.18).

Rabbit production

The rearing of rabbits is a relatively new agricultural enterprise in Sierra Leone. It has, however, a lot of potential, as rabbit meat contains a high amount of quality protein, the rabbits can be cheaply and easily kept and they reproduce rapidly. In addition, rabbits produce valuable by-products in the form of pelts (skins with fur attached), fur and droppings.

Rabbits are cheap and easy to look after. Are you going to try keeping them?

Some key words and terms in this chapter

Livestock Cattle, poultry, horses etc. kept for domestic use but not as pets.

Extensive The keeping of livestock outdoors on a low-cost basis using few resources.

Intensive The keeping of livestock indoors using a high level of resources to achieve a high level of production.

Digestion The process of breaking down and absorbing food in the body.

Dairy A place in which cows are milked or where milk is processed.

Animal husbandry The skill or science of caring for livestock.

Herdsman A person who looks after livestock (a shepherd looks after sheep and a goatherd looks after goats).

Predators Animals that prey (attack or kill) on others.

Pastoral farming Farming that is based around grassland or pasture.

Dairy products Foodstuffs made from milk.

Colostrum The first milk produced by an animal after giving birth. This milk provides protection from disease for the young animal.

Caecum Part of the large intestine. The caecum in rabbits and horses is enlarged to allow the digestion of grass.

Rumen An organ found in cattle, sheep, goats etc., which is used to digest grass and other fibrous foods.

Ruminant An animal with a rumen.

Cellulose An indigestible carbohydrate found in plants.

Crop Found in birds, this storage organ holds food that has been swallowed.

Agricultural inputs Resources required for agricultural production, such as money, labour and fuel.

Gizzard Found in birds, this organ contains grit to grind up food before it is digested and absorbed.

Grit Small fragments of stone that are used by birds to grind up foodstuffs in their gizzard.

Kraaled The process of putting animals in a kraal. A kraal is a pen to hold animals.

Ventilation The supplying of fresh air to a building.

Savanna Open grasslands usually with scattered trees and bushes.

Tilapia A variety of fish which is increasingly being kept in ponds and reservoirs.

Exercises

Multiple choice questions

1 Which of the following is a 'simple stomached' animal?
  a pig
  b cow
  c sheep
  d goat

2 Which of the following animals is a ruminant?
  a chicken
  b goat
  c rabbit
  d pig

3 Which of the following animals has an enlarged caecum to assist in the digestion of its food?
   a sheep   c pig
   b rabbit   d cow

4 Which of the following would you **not** normally associate with intensive livestock production?
   a local breeds
   b high productivity
   c capital investment
   d controlled environment

5 From which neighbouring country do the Fula people come from?
   a Guinea   c Liberia
   b Ivory Coast   d Guinea Bissau

6 A buck and a doe are the names given to the male and female of which domesticated animal?
   a goat   c sheep
   b duck   d rabbit

Missing words

7 Copy the following passage into your exercise book and complete it by selecting the most appropriate word or words from the list below.

Much of the ...... livestock production in Sierra Leone could be described as ...... Our farmers have little ...... to invest in the ...... necessary to ...... the level of production. Each farmer tends to own a few animals, which provide ...... for the family or may be sold to provide cash in an ...... The animals that are kept are usually .... which are hardy and have developed a ...... to many ...... and ...... While the level of .... may not be very high, the system suits our conditions and makes a large contribution to the ....

emergency	**diseases**
traditional	**local breeds**
capital	**raise**
food	**pests**
resources	**national economy**
resistance	**animal husbandry**
extensive	

Crossword puzzle

8 Copy the grid into your exercise book and complete the following clues.

Across
1 Cattle, sheep and goats belong to this group of animals (9)
2 A variety or race of animal (5)
3 These may be a danger to livestock grazing in the bush! (6)
4 The local race of cattle (5)

Down
5 The process of breaking down and absorbing food (9)
1 Animals kept for their meat, fur and pelts (7)
6 The type of fish often raised in ponds or reservoirs (7)
3 The ...... (or hides) of cattle are used to produce leather (5)

Points for discussion

9 What problems would farmers meet if they were to keep their animals intensively?
10 Why do you think that intensive livestock farming may **not** always be profitable?
11 How can we produce more livestock to feed our people and, at the same time, prevent people killing our wildlife for food?
12 How do people market their livestock and livestock products? How could marketing be improved?
13 How are farm animals used or celebrated in traditional ceremonies, songs, dances, theatre and art?

CHAPTER 10 Poultry

Introduction

In this chapter we will look at the basic management, housing, feeding and rearing of poultry; in later books you will learn about poultry husbandry in much more detail. This chapter may encourage you to keep some poultry yourself! Keeping a few hens is an interesting pastime which could help you earn a little money!

What are poultry?

Poultry is a term we have used widely in previous chapters; perhaps it is time we explained what we mean by the word! Poultry is a collective (group) word we use to describe domestic fowl (birds kept for their meat and eggs). We have already discovered the importance of poultry in providing us with meat, eggs and other by-products, such as manure and feathers.

How many different domestic fowl can you name? Make a list of as many as you can think of in your exercise book; your teacher will then make a complete list by asking others in your class.

Poultry terms

Farmers and agriculturalists use a wide variety of terms when referring to poultry. The various terms are used to describe birds of different types, ages and sex. You should try to become familiar with these words.

The term 'fowl' is another group name and means the same as poultry. The term 'chicken' is quite confusing as it is used to refer to hens and cocks, which we keep for eggs and meat; it can also refer to a young hen or cock.

Table 10.1 lists some of the special names for birds of different breeds according to their age and sex.

A group of poultry is usually called a flock; a group of geese may also be called a gaggle.

Table 10.1 Poultry names

Poultry type	Adult male	Adult female	Young male	Young female	Young
chicken	cock	hen	cockerel	pullet	chick(en)
turkey	cock	hen	cock poult	hen poult	chick
duck	drake	duck	— —	— —	duckling
goose	gander	goose	— —	— —	gosling

Reasons for keeping poultry

Let us investigate some of the reasons why poultry keeping is so popular in Sierra Leone. You will know that you do not have to be a farmer to keep poultry; many people who live in towns and villages keep some birds. Why is this? Perhaps before you read the next section you can think of some reasons.

People keep poultry because:

- They may be kept on a small area of land. This makes them suitable for keeping in a backyard or garden.
- They are well adapted to the hot climate of Sierra Leone. The wild ancestors of some of our domestic poultry originated in the tropics.
- They provide us with meat and eggs.
- They grow quickly and money spent on them will soon be repaid through the sale of meat or eggs.
- Keeping chickens of different ages ensures a steady supply of eggs or meat for home consumption or sale.
- They are easy to look after. If they are allowed to wander about freely they can largely fend for themselves.
- If the poultry are kept in an enclosure, their excreta form a useful manure.

Using poultry products

Eggs and egg quality

Eggs can be eaten raw, boiled, fried or used for cooking. Only fresh eggs should be used and, as you will be aware, eggs do not remain fresh for long in a hot climate. Here are three simple exercises to test eggs for freshness.

ACTIVITY 1 Testing eggs for freshness

Test 1 Take several eggs of different ages and put them gently into a bowl of water. Those that stay on the bottom are fresh. Those that float to the top are stale (Figure 10.1).

Test 2 Look carefully at the shells of eggs. Those that have a

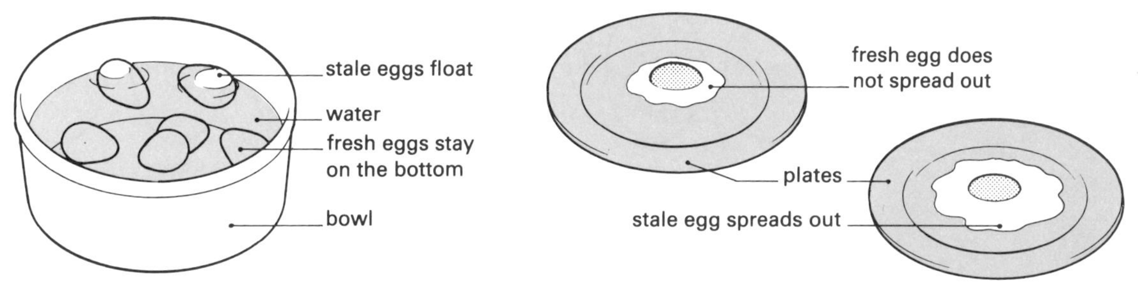

Figure 10.1 Testing eggs for freshness: Test 1 (left)
Figure 10.2 Testing eggs for freshness: Test 3 (right)

rough shell with a matt surface are usually fresh; those that have a smooth shiny shell may be stale. Why do you think this is so?

Test 3 Using eggs from Test 1, break open a fresh egg and a stale egg and put them gently on separate plates. Look carefully at them. You will notice that the fresh egg does not spread out; the yolk and white are strong and hold together, the yolk rises up from the plate (Figure 10.2). The stale egg spreads out much further and the yolk is much flatter.

If you wash eggs, to make them look better or to remove the dirt, they will quickly go bad. If eggs must be cleaned, they should be rubbed gently with a dry cloth.

Chicken meat

Chicken meat is cheap, tasty and healthy; this accounts for its increasing popularity.

When a chicken is killed, the feathers are removed; this is known as 'plucking'. The internal organs (often called the viscera) are removed; this is known as 'drawing'. Some of the organs, such as the heart and liver, may be kept, but the rest are thrown away. When the chicken is ready for cooking it is said to be 'dressed'.

Chickens kept solely for meat production are called 'broilers'. In some countries they are called 'table birds'. Broiling means cooking over an open fire or grill. Young birds of about twelve weeks of age are the best for this as their flesh is tender.

Older birds are tougher and need to be boiled slowly. The chickens sold by local farmers who have small flocks are often over two years old and are of this type!

When you last ate chicken did your mother boil it or broil it? Try not to confuse the meanings of the two words.

Breeding of chickens

Reproduction from eggs

Poultry are not like mammals, which give birth to their young alive; birds lay eggs. If an egg is to hatch (the time at which the

Figure 10.3
The development of a chick embryo

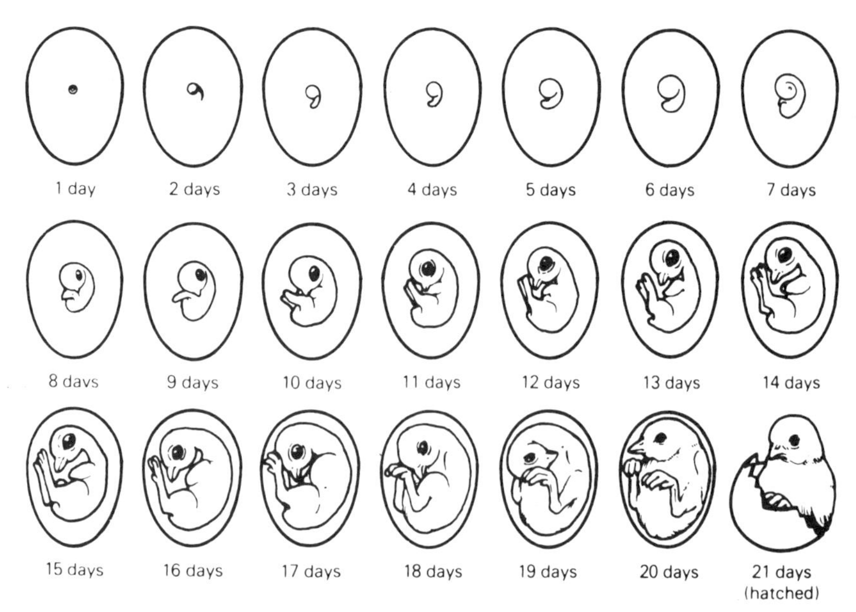

young chick emerges), it must be fertile; this requires a cockerel to have mated with a hen. A fertile egg contains everything needed for the developing embryo (immature chick). Before hatching, an egg must be incubated (kept warm) for a period of some weeks. During this time the food reserves in the yolk and white of the egg are used up to supply the growing chick.

The incubation period for a chicken is 21 days; after this time the chick will peck a small hole in the shell and gradually work its way free (Figure 10.3).

Eggs may be incubated naturally or artificially.

Natural incubation

Over a period of time, a hen will lay several eggs (a clutch). If the eggs are not removed from her she will stop laying and become 'broody'; this means that she is ready to sit on her eggs and incubate them. The hen will not sit on her eggs until she has laid her complete clutch; all the chicks will eventually hatch at about the same time.

The hen will remain on her eggs for most of the incubation period (Figure 10.4). She will only come off her eggs for short periods so that she can eat, and she will make sure that the eggs do not get cold or are exposed to danger. During the incubation period the hen will turn the eggs over several times a day; this is necessary to ensure the proper development of the chicks. Under natural conditions the eggs are kept in a humid (moist) atmosphere; this allows the chicks to develop and to break free from their eggshells.

When the eggs hatch, the chicks are quite wet; they soon dry and are able to move about. Chicks have enough food to last them

Figure 10.4
A broody hen on her nest (left)
Figure 10.5
A free-range hen with her brood of chicks (right)

for the first 24 hours of their lives, so they must quickly learn to find food for themselves. The hen will keep the chicks warm, protect them and help them find food (Figure 10.5). As the chicks grow, the owner may provide them with some additional food. Very young chicks love boiled egg yolk! As they grow older they can be fed special chick mash or meal, broken grain, bran or household waste.

Artificial incubation

Artificial incubation is the process of incubating eggs without the help of hens. This allows farmers to raise large numbers of chicks

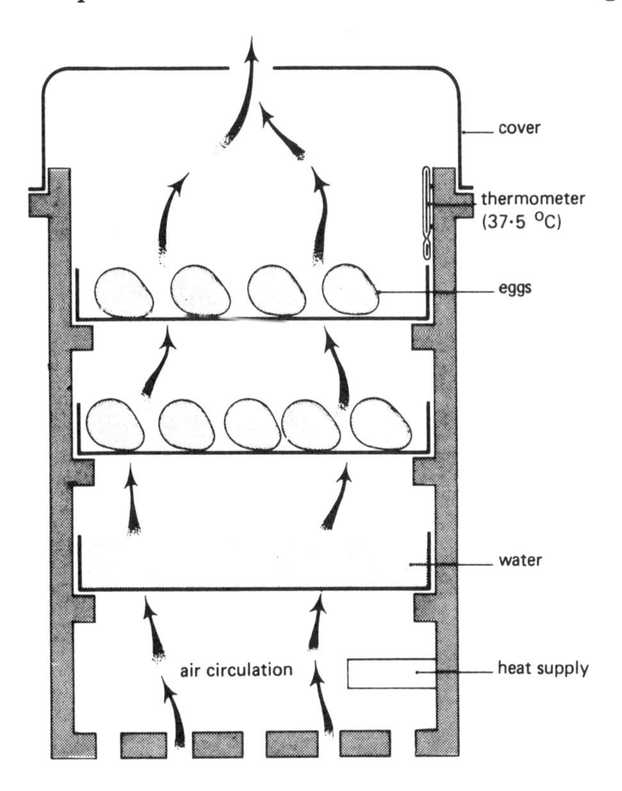

Figure 10.6
An artificial incubator

Figure 10.7
A simple artificial brooder: the heat is provided by an overhead lamp

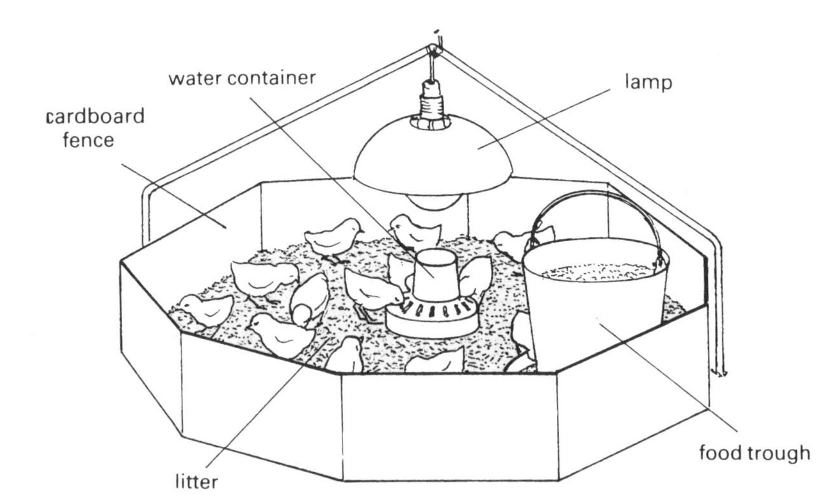

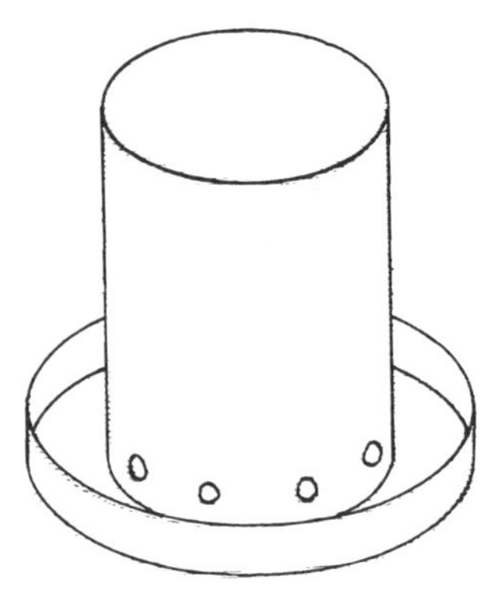

(a) drinker

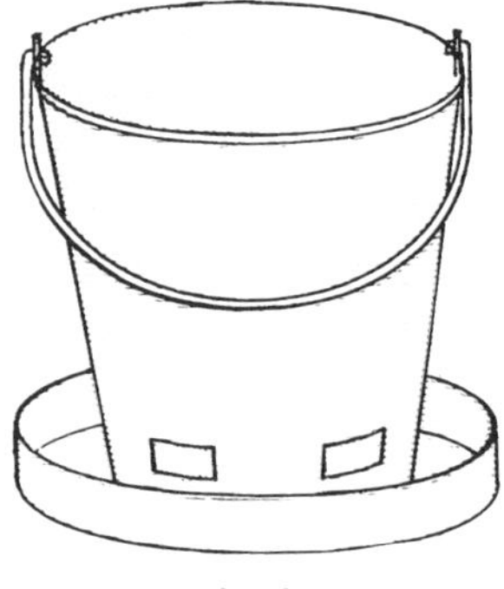

(b) feeder

at one time. It means that our broiler and laying industries can have the necessary regular supplies of birds all of the same age.

Eggs are placed on trays in special man-made incubators of the type shown in Figure 10.6. Temperature and moisture in the incubator are carefully controlled. In small incubators, the temperature may be controlled by a paraffin heater and the humidity is maintained by topping up a small water bath. In very large incubators, temperature and humidity are controlled electrically.

The eggs are moved several times a day to ensure that the chicks do not die; in the smaller machines this is done by hand; in the larger machines it is carried out mechanically.

Once the chicks have hatched they are removed from the incubator and are reared artificially.

When large numbers of chicks have been hatched in an incubator they must be looked after; this is a skilled job. First, the chicks may have to be sexed, when the males and females are separated; this is, of course, very necessary if the chicks are to be raised for egg laying! The chicks have to be fed on a carefully controlled diet given in special feeders; they have to be kept warm under brooders (artificial heaters) (Figure 10.7).

Once the chicks are big enough they may be moved into the laying or fattening houses.

Breeds of chicken

Over time, people have developed many different breeds of poultry, each one having its own special characteristics.

Local chickens

The chickens you see in our villages are often a mixture of breeds and are rather small. Our local chickens are quite hardy and resistant to disease; they grow slowly and produce a few small eggs.

Breeds from foreign countries

Many attempts have been made to improve our poultry, and breeds of chickens have been brought in from abroad; these are called exotic breeds (Figure 10.8).

Exotic breeds are generally egg producers or meat producers; some birds are called dual-purpose breeds, as they are kept for both eggs and meat. Exotic breeds are usually larger than our local chickens and these bigger birds require more food and a better level of management.

Below is a description of three imported breeds of chickens:

White Leghorn. These are laying birds. They can lay up to 200 eggs per year. The birds are rather nervous and excitable.

Light Sussex. These are table birds. They grow quickly and are ready for eating in twelve weeks.

Rhode Island Red. These are dual-purpose birds. As you have probably guessed, they lay many eggs and also put on a lot of meat, so they make good table birds as well.

Do you know of anyone who keeps any exotic breeds of chickens? What breeds do they keep? What do they look like? Are they kept for eggs or meat?

Hybrids

A hybrid is the name given to an animal (or plant) resulting from a cross between genetically unlike individuals (different species or varieties).

Hybrid poultry are often very productive. The use of hybrids is very common in the intensive poultry industry where the production of layers and broilers is a very specialised business.

It is possible for us to produce hybrid chickens by mating our local hens with a cockerel of an imported breed. With luck, the best of both breeds will be seen in the offspring! They should lay more eggs, which should be larger; the birds should grow bigger and faster than our local chickens; and they should be able to adapt to village conditions.

Figure 10.8 Exotic breeds of chicken

Light Sussex

Rhode Island Red

Do you know of anyone who has tried to improve local chickens by mating them with an exotic breed? Did it work? What did the resulting chickens look like?

Feeding chickens

The feeding of chickens will largely depend on how the birds are kept. We find a great difference between the feeding of birds under our traditional system as compared with feeding in the intensive layer or broiler industries.

Figure 10.9 Giving local poultry extra feed will increase their egg production

Whichever system is in operation, the farmer should provide the birds with a balanced diet which will keep them healthy and allow them to produce eggs or meat to their full capacity (Figure 10.9).

We have seen that, under our traditional system of poultry keeping, the birds must largely fend for themselves. By moving about freely, the birds are able to enjoy a varied, but meagre, diet. The birds will pick up insects, worms and other living creatures and will eat fruit and pieces of green plants; they will scratch around for grit. The farmer may provide the chickens with some additional feed.

In intensive poultry keeping, feeding is an exact science. The birds are kept in very large numbers and have no access to any other form of food. Everything the birds require must be supplied in their feed. Feed is so expensive that the amount of food given to the birds must be closely regulated. Poultry farmers will buy specially prepared, complete rations from a feed mill. There are special feeds for growing chickens, layers and broilers.

We must not forget that poultry need a good supply of water. As for all livestock, this water should be pure.

Poultry management

We are already aware that poultry may be kept in many different ways. It is time for us to consider some of these in more detail and to look at some of the buildings and equipment required for each system of production. We can say that poultry can be kept under three basic systems; these are:

- Extensive systems, sometimes called free-range systems (where the chickens are free to move around outside).
- Semi-intensive systems (where there is a chicken house and an enclosed piece of land).
- Intensive systems (where chickens spend all their lives inside a building).

Extensive systems

We are already familiar with this, our traditional system of poultry keeping. We know that this system uses local breeds of birds, which are kept free-range, and that there is little or no investment in feed, buildings or equipment. We would expect production to be low from this system of poultry keeping.

Some of the drawbacks of our traditional extensive system can be overcome by exercising some control over the chickens. A farmer may provide a small wooden house, which is moved to areas where food (such as crop residues) is available; the house may be put on wheels so it is easier to move. If nesting boxes are built into the house, the hens will lay in them and egg collection is easier. The birds are shut up in the house at night for protection. This system, however, requires more work.

Figure 10.10
A semi-intensive poultry rearing system

Semi-intensive systems

'Semi' means half or partly, and these systems, as you might expect, are halfway between extensive and intensive systems! These systems might be very suitable for Sierra Leone; they would not need a great deal of money and farmers would be assured of higher production (Figure 10.10). We will consider two examples.

Chicken house with a run

In this system, all that is needed is a chicken house surrounded by a chicken run (an area enclosed by a high, wire, chicken-proof fence). The chickens are safe from their enemies outside! The chickens are not free to wander about so the farmer must supply all their food and water. It is desirable that a farmer moves the chicken house and run to a new location every year; this stops chicken diseases building up in the soil (Figures 10.11 and 10.12).

Figure 10.11
A chicken house with a run

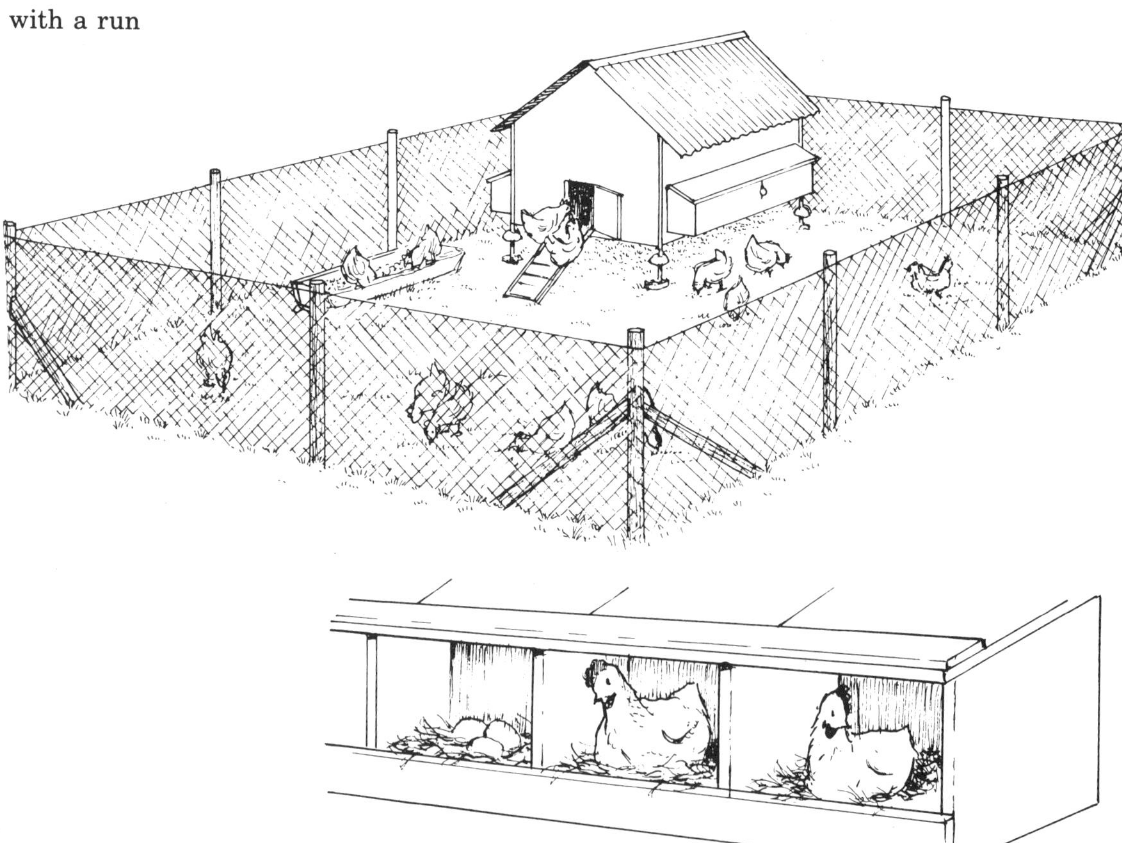

Figure 10.12
In a chicken house, the hens learn to lay their eggs in the nest boxes

Ark

If a farmer owns a small number of chickens, an 'ark' (or 'fold') may be used (Figure 10.13). The ark is a small house with wheels, nesting boxes and no floor. The birds are able to feed on the area under the house. The ark is moved to a new place each day. The farmer will still have to give the chickens a lot of extra food and provide water.

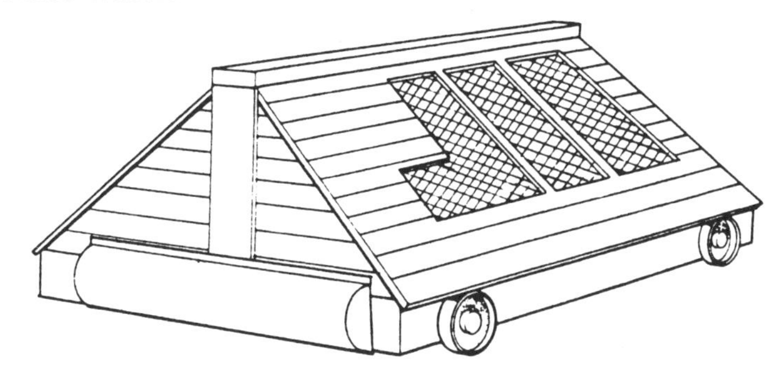

Figure 10.13
A portable fold unit is moved to a new position every day

Intensive systems

Very large flocks of poultry can be kept under these systems. Flocks of birds are kept in large permanent houses and spend all their lives inside.

We can identify two different intensive systems:

- deep litter
- battery cages

Deep litter system

The deep litter system is suitable for both broiler and egg production. Here the poultry are given the freedom to move around inside the poultry shed (Figure 10.14).

Figure 10.14
A deep litter system with nest boxes, perches, feeders and drinkers

Under this system, broiler production takes about 10 weeks; when the birds reach a given weight they are slaughtered and the house is restocked.

When laying hens are kept on deep litter, nest boxes have to be provided. By giving the birds perches to roost (sleep) on, farmers are able to keep more birds in one house.

The floor of the house is covered with litter consisting of a dry vegetative material, such as rice straw, dried grass, groundnut shells or wood shavings (saw dust is not used for health reasons). New litter should be added each day and allowed to build up. The litter acts like a compost heap and the bird droppings are quickly broken down; there should be no smell and the birds should remain healthy. When the house is empty of birds the litter is dug out and used as manure.

Battery cage system

This is a system which is suitable for large-scale egg production. Special hybrid hens are used as they can adapt to this intensive system. It is fair to say that many people do not approve of this method of poultry keeping; they feel it is cruel to keep the birds in this way (Figure 10.15).

When the hens are about to lay, they are put in small cages with about five birds to each cage. The chickens will spend the rest of their lives in here. The cages are arranged in rows and are stacked on top of each other; cages may be up to five rows high.

The cages have wire floors so that the droppings may fall through on to a belt or trays below. The droppings can then be removed without opening the cage doors. Food, in the form of a ground up meal, is always available to the hens. Water is supplied to each cage through small taps called 'nipple drinkers'.

Egg production in hens depends on the amount of daylight they receive. The light in a battery house has to be artificially controlled and electric lights are turned on for a certain length of time each day.

The eggs roll down a sloping floor away from the hens so they cannot be pecked or damaged. They may then be removed for grading (sorting into different sizes and qualities) and packing (putting into boxes or trays for sale) (Figure 10.16).

The houses must be carefully ventilated and electric fans are used to draw out the stale air.

With such a concentration of birds, it is important that they remain healthy. Their health should be monitored each day and diseased birds must be taken away immediately.

Some advantages of intensive systems

Let us consider some of the advantages of keeping poultry in deep litter houses or battery cages:

Figure 10.15
A battery hen house (left)

Figure 10.16
In a battery hen house, the egg passes through the bars and on to a sloping floor. The hen cannot then damage it and it can easily be collected by the farmer (right)

- Very little land is required and one can keep a large number of birds.
- The chickens get a balanced diet; each gets the correct quantity and quality of food. The birds grow quickly and the hens lay plenty of eggs.
- The birds' environment can be carefully controlled. The correct lighting, temperature and ventilation can help to get maximum production.
- The manure is a valuable fertiliser and is easily collected from inside the deep litter house or under the battery cage.

Some disadvantages of intensive systems

- Intensive systems are very expensive ways of keeping chickens. The initial cost (buildings, equipment etc.) of setting up these enterprises is very high. The running costs (food, electricity, maintenance etc.) are also considerable.
- A high degree of management expertise is needed. It is also necessary to employ skilled workers who will be on hand at all times.
- The site must be near a main road and a market. Reliable transport must be available.
- An emergency generator must be able to provide electricity in case of power failure.
- Diseases must be treated quickly or the entire flock may be wiped out.

Health in poultry

Success in keeping poultry depends, as in so many other branches of livestock husbandry, on keeping the animals healthy. If you are going to try and keep a few birds you must make sure that you start off with healthy stock (Figure 10.17). Let us consider what we should look for in a healthy bird:

Figure 10.17
A healthy bird

- The bird should look alert and have bright eyes with no gummy substances around them.
- It should have a wattle and comb that are large and bright.
- It should have clean and shiny feathers, which do not stick out from its body.
- The bird should be able to move about normally taking an interest in its surroundings while searching for food.

If you are going to buy some chickens from a farmer, look at them carefully and remember these four points before you pay for them!

Looking after your poultry

When you have bought your poultry you will want to care for them in the best possible way. Here are a few things you should try to do:

- Provide them with a regular supply of good food and clean water each day.
- Provide shelter from rain and hot sunshine.
- Allow them to have an area of land to scratch about on; the birds should be given a fresh area from time to time.
- Keep their house clean.
- Keep them safe from predators.
- If a bird shows signs of illness, get rid of it quickly. Birds that recover from an illness seldom do well again.
- Dead birds should be taken away at once and burned; they should not be eaten.

If you look after your poultry, they will reward you with meat and eggs.

Some key words and terms in this chapter

Poultry A collective (group) word describing all birds kept for domestic farm use.
Fowl A word that means the same as poultry.
Chickens A word meaning hens or cocks (or the young of several different types of poultry).

Poult An immature bird.
Plucking The removal of the feathers from a dead bird.
Viscera The internal organs.
Drawing The removal of the internal organs.
Dressing The preparation of a bird or other animal for cooking.
Broiler A chicken reared for meat production. Broilers are usually killed at or before 12 weeks of age. These birds are tender and are suitable for roasting or grilling.
Table bird Another term for a broiler.
Hatching The emergence of chicks from their eggs.
Incubate The process of keeping eggs warm for a period of time to allow the embryos to develop.
Incubation period The length of time that eggs must be kept warm to allow the chicks to develop. In chickens the incubation period is 21 days.
Clutch A batch of eggs incubated by a bird.
Broody A term that describes a bird sitting on a clutch of eggs.
Incubator A machine used to incubate eggs.
Brooder A special heater which is used to keep chicks warm in intensive poultry production; a brooder takes the place of a hen.
Exotic breed A term used to describe a breed which has been brought into a country from outside.
Dual-purpose breed A breed of animal kept for two purposes. For example, Rhode Island Red poultry are kept for meat and egg production.
Hybrid A name given to an animal (or plant) produced by crossing two genetically unlike individuals. The resulting cross often has useful characteristics.
Chicken run An enclosed area in which poultry may move about freely and in which they are protected from predators.
Ark A small mobile chicken house and run. It may sometimes be called a fold.
Deep litter An intensive system of poultry keeping where the birds are kept in a building and allowed to move about freely on a floor covered with a litter of wood shavings or similar material. Deep litter systems are suitable for meat or egg production.
Battery cages An intensive system of poultry keeping in which chickens are kept in cages to produce eggs.
Nest box A special box provided for birds to lay their eggs in.
Perches Wooden bars or poles provided for chickens to roost (sleep) on.
Grading The sorting of eggs into different sizes.
Packing Putting eggs into boxes, cartons or trays for transport and sale.

Exercises

Multiple choice questions

Write the correct answer in your exercise book.

1 Which of the following statements is true? Poultry kept in an extensive system
   a do not occupy much land
   b are very expensive to maintain
   c lay large numbers of eggs
   d are hardy and resistant to disease

2 To keep chickens healthy they
   a should be left to look after themselves
   b must find their own shelter
   c need regular feeding
   d do not require water

3 Deep litter
  a is found on the floor of battery cages
  b consists of saw dust
  c is a group of new born piglets
  d is a system of broiler production

4 Which of the following breeds is kept for egg production only?
  a Rhode Island Red
  b White Leghorn
  c Red Sussex
  d local breeds

5 When fresh eggs are placed in a bowl of water they
  a sink and then rise to the surface
  b float on the top
  c sink to the bottom
  d float for a while then sink

6 Pulling out the feathers from a chicken before cooking is called
  a plucking
  b drawing
  c trussing
  d dressing

7 The name for an adult male chicken is
  a cock
  b pullet
  c drake
  d gander

8 Hens eggs hatch in
  a 7 days
  b 14 days
  c 21 days
  d 28 days

Missing words

9 Copy the following passage into your exercise book and complete it by selecting the most appropriate word or words from the passage below.

A hen may select a quiet spot in which to make her ...... When she has laid a number of ...... she will become ...... She will then ...... her ...... During this time she will ...... her eggs and keep them ...... After the incubation ...... of ...., the eggs will ...... The ...... should all emerge at the same time and will quickly start to feed.

period	**eggs**	**twenty-one days**
broody	**chicks**	**hatch**
turn	**nest**	**clutch**
warm	**incubate**	

Crossword puzzle

10 Copy the diagram into your exercise book and complete the following clues.

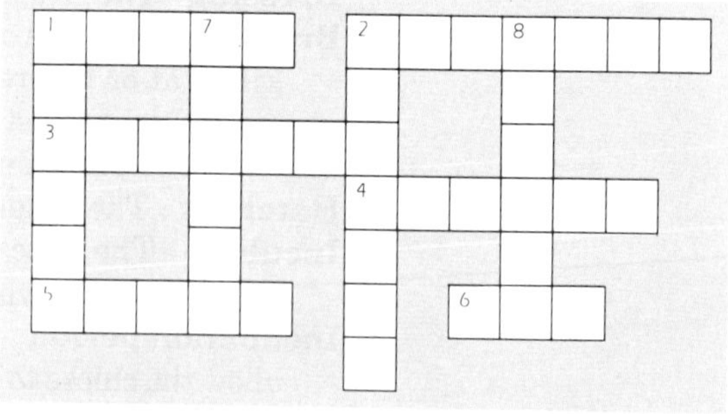

Across
1 We say that eggs do this when the chicks break out from the eggs in which they have been incubated (5)
2 A word used to describe all domesticated birds (7)
3 A chicken reared for meat and suitable for grilling or roasting (7)
4 A batch of eggs incubated by a hen (6)
5 To prepare a bird for eating (5)
6 A small mobile chicken house and run (3)

Down
1 A bird produced by the mating of two distinctly different types (6)
2 Poles provided for poultry to roost on (7)
7 Very young cocks or hens (6)
8 Deep ...... is a system of intensive poultry production (6)

Points for discussion

11 Do you think that keeping poultry in battery cages or deep litter systems is cruel? Does our need to produce cheap eggs and meat justify this method of production?
12 What problems might we find if we try to make our local poultry keeping more productive? How would you try to improve our traditional system of poultry keeping?
13 How are poultry used in the traditional life of our villages. Do you know of any songs, dances or stories that involve poultry?

GLOSSARY

A list of English, Mende, and Temne names of common trees, crops and wild flowers found in Sierra Leone.

English	Mende	Temne
bambara nut	manikri	nikri
bamboo	nduvui	e-pote
		ka-sule
carpet grass	yani	ka-gbatha
	ndiwi	
cashew nut	kundi	e-lil-epoto
cassava	tangei	a-yoka
citrus	dumbele	a-lambray
cocoa	coqui	a-coco
coconut	pu-lolui	angbara-poto
		koe-b'ara
cocoyam	kokoi	e-koko
coffee	yonembi	a-coffi
cola (kola)	tui	ang-ola
	tolui	a-korla
cotton	fande	gbonk
	fande-wuli	e-pompo
cowpea	kondoi	ka-beant
elephant grass	ngali	awo
groundnut	nyui	a-kan
	nikili	
imperata grass	teli	ka-lat
mahogany	bafili	yeamani
maize	nyoi	koe-mank
millet	gbeli-nyoi	ta-sur
mango	mangoi	a-mangoro
oil palm	tokpoi	a-komp
okra	bende orbondo	a-lontho
onion	yabas	ta-yaba
peppers	pujei	e-bembe
		e-gbemgbe
piassava	kajo	ma-yoi
raphia	duvui	a-kentr
rice	mbei	koe-yaka
sekou touri leaves	seku turay daway	e-bopre ya-seku thuray
sensitive plant	kpete nana	ka-bomopneh
silk cotton tree	nguwe	a-polon
simsim (beniseed)	mande	c-yente
sisal	mande	e-yente
sorghum	keti	tag boyo
soya bean	torwai	a-lale
	(torwawa)	
sugar cane	nyokoi	koe-gbokang
sunflower	flaway	a-flawa
sweet potato	jowoi	a-muna
	njowoi	

tobacco	tawa	taba
tomato	ki-bongie	ma-thamba
water leaf (Lagos bologi)	kembe	a-kempa
water yam	mbui	eyams
yams (various)	foli	e-bunk
	kpuli	am-bunk
	mboli	

ANSWERS TO EXERCISES

Chapter 1
Activity 6: 1 (c) 2 (d) 3 (a) 4 (a) 5 (d) 6 (b)
1 erosion 2 firewood 3 exploitation
4 conservation 5 desert
Exercises Puzzles: 1 Across: 1 cattle 2 pigs
Down: 3 rabbits
2 Across: 1 north 2 millet Down: 3 rain 4 rat
3 (a) rice (b) kid (c) cow (d) cotton
Multiple choice: 4 c) 5 (b) 6 (c) 7 (d) 8 (b)
9 (c) 10 (d)
Matching items: 11 (a) raw materials (b) foreign exchange (c) food and clothing (d) income and employment
12 (a) (viii) (b) (vii) (c) (vi) (d) (i) (e) (ii)
(f) (iii) (g) (iv) (h) (v)

Chapter 2
Multiple choice: 1 (c) 2 (a) 3 (c) 4 (b) 5 (d)
6 (b)
Puzzles: 7 water loam
8 Across: 1 organic 2 slippery 3 horizon 4 silt
Down: 5 transported 6 profile
Missing words: 9 (i) rainfall (ii) moisture (iii) wilt
(iv) irrigate (v) pore space (vi) drains
(vii) particles (viii) roots

Chapter 3
Multiple choice: 1 (c) 2 (b) 3 (a) 4 (c) 5 (a)
6 (c) 7 (d) 8 (c) 9 (a)
Matching items: 10 (a) broadcasting (b) pocket method
(c) top dressing (d) split dressing (e) green dressing
Missing words: 11 (i) nutrients (ii) roots (iii) stem
(iv) carbon dioxide (v) photosynthesis (vi) vessels
Crossword puzzle: 12 Across: 1 powder 2 leaching
Down: 1 prills 3 ring 4 water 5 humus

Chapter 4
Word puzzle: 1 hoes 2 fork 3 rake 4 line
Missing words: (a) hand trowel (b) rose watering can
(c) fork (d) wheelbarrow front (e) cleans oils
(f) dangerous sharp
Crossword: 3 Across: 1 spades 2 garden line
3 rivets Down: 1 secateurs 4 scales 5 rake
6 axe
Multiple choice: 4 (a) 5 (c) 6 (a) 7 (a) 8 (b)
9 (b) 10 (a) 11 (d) 12 (a) 13 (c) 14 (a)
15 (b) 16 (c)

Chapter 5
Missing words: 1 (a) water (b) heat sun (c) wind
(d) tractor (e) ox living (f) strength
Multiple choice: 2 (b) 3 (c) 4 (c) 5 (b) 6 (b)
Crossword puzzle: 7 Across: 1 methane 2 solar power
3 ox 4 power 5 coal Down: 6 water wheel
7 fuel 8 tractor 9 work

Chapter 6
Multiple choice: 1 (d) 2 (c) 3 (c) 4 (d) 5 (c)
Missing words: 6 (i) annuals (ii) perennials
(iii) seeds (iv) reproduced (v) vegetative propagation
(vi) quality (vii) germination percentage (viii) pests
Matching items: 7 (a) iii (b) v (c) ii (d) iv (e) i
Crossword puzzle: 8 Across: 1 layering 2 onion
3 legume 4 biennial Down: 5 annual 6 runner
7 scion 8 bulb

Chapter 7
Multiple choice: 1 (d) 2 (d) 3 (a) 4 (a) 5 (a)
6 (b) 7 (b) 8 (c)
Missing words: 9 (i) feed (ii) risk (iii) failure
(iv) spread (v) pests (vi) workload (vii) harvested
(viii) surplus (ix) cash
Crossword puzzle: 10 Across: 1 sorghum 2 fibre
3 piassava 4 acacia 5 orange Down: 1 sweet potato 6 raffia 7 bass 8 yams

Chapter 8
Multiple choice: 1 (c) 2 (d) 3 (b) 4 (d) 5 (b)
6 (a)
Missing words: 7 (i) seeds (ii) vegetables
(iii) agricultural merchant (iv) instructions
(v) mature (vi) harvested (vii) marketed
(viii) quality (ix) profit
Crossword puzzle: 8 Across: 1 stones 2 damping off
3 seed 4 mulch 5 watering 6 tomato
Down: 7 transplanting 8 germinate 9 cabbage
10 pests

Chapter 9
Multiple choice: 1 (a) 2 (b) 3 (b) 4 (a) 5 (a)
6 (d)
Missing words: 7 (i) traditional (ii) extensive
(iii) capital (iv) resources (v) raise (vi) food
(vii) emergency (viii) local breeds (ix) resistance
(x) pests (xi) diseases (xii) animal husbandry
(xiii) national economy
Crossword puzzle: 8 Across: 1 ruminants 2 breed
3 snakes 4 N'dama Down: 5 digestion 1 rabbits
6 tilapia 3 skins

Chapter 10
Multiple choice: 1 (d) 2 (c) 3 (d) 4 (b) 5 (c)
6 (a) 7 (a) 8 (c)
Missing words: 9 (i) nest (ii) eggs (iii) broody
(iv) incubate (v) clutch (vi) turn (vii) warm
(viii) period (ix) twenty-one days (x) hatch
(xi) chicks
Crossword puzzle: 10 Across: 1 hatch 2 poultry
3 broiler 4 clutch 5 dress 6 ark
Down: 1 hybrid 2 perches 7 chicks 8 litter